STUDENT UNIT GUIDE

NEW EDITION

OCR AS Geography Unit F761
Managing Physical Environments

Michael Raw

With thanks to Katrina Hillary, Sarah Whitehead, Henry Briggs and Joshua Brear

Philip Allan Updates, an imprint of Hodder Education, an Hachette UK company, Market Place, Deddington, Oxfordshire OX15 0SE

Orders
Bookpoint Ltd, 130 Milton Park, Abingdon, Oxfordshire OX14 4SB
tel: 01235 827827
fax: 01235 400401
e-mail: education@bookpoint.co.uk
Lines are open 9.00 a.m.–5.00 p.m., Monday to Saturday, with a 24-hour message answering service.
You can also order through the Philip Allan Updates website: www.philipallan.co.uk

ISBN 978-1-4441-7179-2

First printed 2012
Impression number 5 4 3 2
Year 2016 2015 2014 2013

Cover photo: Artman/Fotolia

Typeset by Integra Software Services Pvt. Ltd., Pondicherry, India

Printed in Dubai

P2087

Contents

Content Guidance

Questions and Answers

Getting the most from this book

Examiner tips

Advice from the examiner on key points in the text to help you learn and recall unit content, avoid pitfalls, and polish your exam technique in order to boost your grade.

Knowledge check

Rapid-fire questions throughout the Content Guidance section to check your understanding.

Knowledge check answers

1 Turn to the back of the book for the Knowledge check answers.

Summary

Summaries

- Each core topic is rounded off by a bullet-list summary for quick-check reference of what you need to know.

Questions & Answers

Exam-style questions

Examiner comments on the questions

Tips on what you need to do to gain full marks, indicated by the icon ⓔ.

Sample student answers

Practise the questions, then look at the student answers that follow each set of questions.

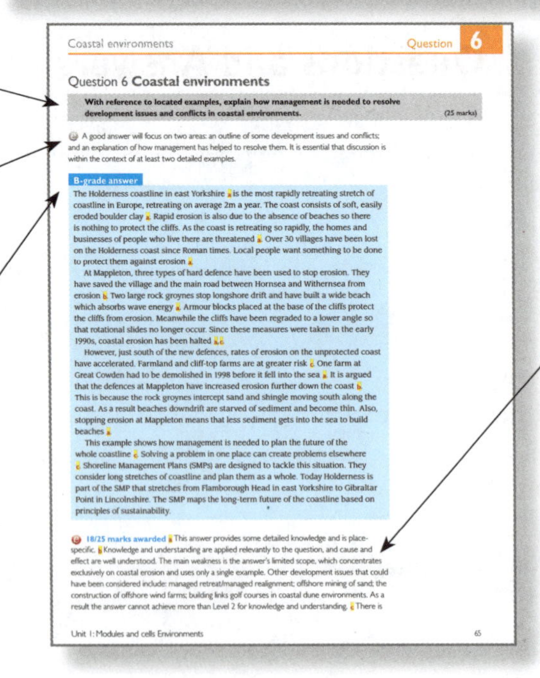

Coastal environments — Question 6

Question 6 **Coastal environments**

With reference to located examples, explain how management is needed to resolve development issues and conflicts in coastal environments. (25 marks)

ⓔ A good answer will focus on two areas: an outline of some development issues and conflicts; and an explanation of how management has helped to resolve them. It is essential that discussion is within the context of at least two detailed examples.

B-grade answer

The Holderness coastline in east Yorkshire a is the most rapidly retreating stretch of coastline in Europe, retreating on average 2m a year. The coast consists of soft, easily eroded boulder clay a Rapid erosion is also due to the absence of beaches so there is nothing to protect the cliffs. As the coast is retreating so rapidly, the homes and businesses of people who live there are threatened a Over 30 villages have been lost on the Holderness coast since Roman times. Local people want something to be done to protect them against erosion a

At Mappleton, three types of hard defence have been used to stop erosion. They have saved the village and the main road between Hornsea and Withernsea from erosion b Two large rock groynes stop longshore drift and have built a wide beach which absorbs wave energy a Armour blocks placed at the base of the cliffs protect the cliffs from erosion. Meanwhile the cliffs have been regraded to a lower angle so that rotational slides no longer occur. Since these measures were taken in the early 1990s, coastal erosion has been halted a,c

However, just south of the new defences, rates of erosion on the unprotected coast have accelerated. Farmland and cliff-top farms are at greater risk c One farm at Great Cowden had to be demolished in 1998 before it fell into the sea a It is argued that the defences at Mappleton have increased erosion further down the coast b This is because the rock groynes intercept sand and shingle moving south along the coast. As a result beaches downdrift are starved of sediment and become thin. Also, stopping erosion at Mappleton means that less sediment gets into the sea to build beaches a

This example shows how management is needed to plan the future of the whole coastline c Solving a problem in one place can create problems elsewhere c Shoreline Management Plans (SMPs) are designed to tackle this situation. They consider long stretches of coastline and plan them as a whole. Today Holderness is part of the SMP that stretches from Flamborough Head in east Yorkshire to Gibraltar Point in Lincolnshire. The SMP maps the long-term future of the coastline based on principles of sustainability.

ⓔ **18/25 marks awarded** a This answer provides some detailed knowledge and is place-specific. b Knowledge and understanding are applied relevantly to the question, and cause and effect are well understood. The main weakness is the answer's limited scope, which concentrates exclusively on coastal erosion and uses only a single example. Other development issues that could have been considered include: managed retreat/managed realignment; offshore mining of sand; the construction of offshore wind farms; building links golf courses in coastal dune environments. As a result the answer cannot achieve more than Level 2 for knowledge and understanding. c There is

Unit 1: Modules and cells Environments 65

Examiner commentary on sample student answers

Find out how many marks each answer would be awarded in the exam and then read the examiner comments (preceded by the icon ⓔ) following each student answer. Annotations that link back to points made in the student answers show exactly how and where marks are gained or lost.

About this book

This guide is designed to help you prepare for OCR AS Geography **Unit F761: Managing Physical Environments**.

The **Content Guidance** section outlines the specification content and the key themes used to formulate examination questions.

The **Questions and Answers** section explains the assessment structure and outlines the techniques for dealing with data-response and extended-writing questions. It provides eight specimen questions (four data-response questions and four extended-writing questions) and two student answers for each question, ranging from grade A to grade E. Each student answer is followed by examiner's comments.

Content Guidance

This section provides a summary of the key ideas and content detail needed for AS Geography Unit F761: Managing Physical Environments.

The content is divided into four main areas:

- **River environments**
- **Coastal environments**
- **Cold environments**
- **Hot arid and semi-arid environments**

When you revise it is important to use a framework that reflects how examiners might test your knowledge and understanding. Therefore, in addition to the key ideas and content detail, this section provides key questions and answers for each topic.

You should study the questions and answers carefully and organise your revision around them. Focusing on the key questions and adding details of your own to the answers should give you a head start in the final examination.

It is essential that you learn the terminology used in the answers to the key questions, particularly the words in **bold** type. You must be confident in applying these terms appropriately in your exam answers.

River environments

What processes and factors are responsible for distinctive fluvial landforms?

Key ideas	Content
Slope processes and channel processes give rise to distinctive fluvial landforms	The study of a river basin or river basins to illustrate: • a range of features associated with erosion in river systems, e.g. meanders, valleys, floodplains, terraces • a range of features associated with deposition in river systems, e.g. bars, alluvial fans, levées, deltas, estuaries • the factors affecting the development of these features including rock type and structure, slope, climate and sea-level change • the processes responsible for these features, including weathering, mass movement, erosion and deposition
Fluvial and sub-aerial processes are influenced by a range of factors that vary from place to place	

Key questions

How do weathering and slope processes influence the development of fluvial landforms?

Weathering and slope processes (known collectively as **sub-aerial processes**) influence fluvial landforms in two ways. First, they are a major source of sediment input to streams and rivers, contributing to the **bedload** and **suspended load**. Deposition and **aggradation** (deposition of sediment within stream and river channels) cause the sediment loads to form features such as **channel bars** and **floodplains**. Second, sub-aerial processes such as **physical** and **chemical weathering**, **surface wash**, **landslides** and **mudflows** cause the **backwasting** of valley slopes, the widening of valleys and the lowering of slope angles. The processes also contribute to bank collapse due to undercutting (by fluvial erosion) and slumping (due to gravity).

Examiner tip The difference between sub-aerial processes (i.e. weathering and slope processes) and fluvial processes (i.e. river erosion, transport and deposition) must be clearly understood. Relevant answers to this question will focus on sub-aerial processes with only passing reference to fluvial processes.

How do weathering and slope processes contribute to a river's load?

Physical and chemical weathering break up exposed rocks at or near the surface. **Surface wash** may then transport smaller particles into streams and rivers, where they form part of the bedload and suspended load. Larger rock fragments on steep slopes above a river channel, weathered by **freeze–thaw**, may roll or slide into the channel to become part of the bedload. Chemical processes may dissolve minerals that are transferred in solution to stream and river channels to form the **solution load**.

Slope processes operate on valley-side slopes through gravity. **Soil creep** transports fine particles slowly downslope. Rapid transfer may result from landslides or **rotational slides** and dramatic slope collapse, often triggered by a stream or river undercutting a slope. All of these mass movement processes transfer rock debris into stream and river channels and contribute to a river's load.

Examiner tip In addition to weathering and slope processes, don't forget that fluvial erosion also contributes directly to a river's load. Also remember that fluvial erosion often leads to channel bank and valley slope instability, which triggers collapse and mass movement.

How do fluvial and sub-aerial processes create distinctive fluvial landforms?

In this question we will consider one landform — V-shaped upland river valleys — though the same approach could be applied to other fluvial landforms such as waterfalls, floodplains, terraces, point bars and so on.

Knowledge check 1 What are the main features of a river's bedload?

V-shaped valleys are formed by a combination of fluvial and sub-aerial processes (see Figure 1 on p. 8).

Upland rivers have steep gradients and a lot of surplus energy. Most **surplus energy** available for erosion acts vertically. **Abrasion** of a channel bed by the transport of coarse bedload particles deepens valleys. Where the channel meets the valley side, undercutting leads to instability, slope collapse and valley widening.

Sub-aerial processes such as freeze–thaw weathering, soil creep, surface-wash and slumping cause valley slopes to backwaste and create typical slope angles of 30–50°. The outcome of fluvial and sub-aerial processes is an open, V-shaped valley.

Examiner tip Questions that require explanations of landforms must focus on the connections between processes (i.e. erosion, deposition, weathering, mass movement) and form (e.g. scale, shape in plan and profile).

Figure 1 The formation of V-shaped valleys

How do rock type, structure, slope, climate and sea-level change influence the development of fluvial features?

When river channels cross bands of resistant rock (which erode slowly) they form waterfalls and rapids. High Force waterfall on the River Tees has formed where a resistant band of **dolerite** crops out in the channel. Fluvial erosion on more resistant rocks is likely to result in relatively narrow and steep-sided valleys. **Rock structure** includes **jointing**, **faulting** and the **bedding** of rocks. River courses may follow fault lines. Channel erosion can expose solid rock that, if bedded horizontally, may form a series of channel steps. Resistant rocks that break up into blocky particles can give rise to particularly coarse bedload with broad, shallow channels and numerous channel bars.

Steep slopes give rivers more energy, promoting **channel incision** through vertical erosion and the formation of V-shaped valleys and **interlocking spurs**. Gentle slopes mean rivers have shallow gradients that favour channel deposition.

Arid and semi-arid climates limit weathering and mass movement processes. As a result, many river valleys have gorge or canyon-like shapes in cross section. If the climate becomes wetter and river discharge increases, renewed downcutting by rivers can produce **river terraces** and **incised meanders**. Similar features form when the sea level falls during a glacial period or the land level rises due to tectonic movement.

Knowledge check 2

What are the main differences between meandering and braided river channels?

In what ways can river basins be a multi-use resource?

Key ideas	Content
River landscapes provide opportunities for a number of human activities, including: industrial development, transport, residential development, energy development, water supply, recreation and leisure, conservation	The study of at least two contrasting river environments to illustrate: • the range of activities found in these river basins, e.g. HEP, water transport, tourism, water supply, recreation and leisure • the reasons for the growth and development of these activities, e.g. population growth, development of agriculture and other economic activities, water shortages • how differing land uses may conflict in these areas, e.g. wildlife and water extraction, conservation and dam building, displacement of populations by dam building

Key questions

In what ways are river basins multi-use resources?

Many river basins have multiple uses. They:

- supply water for domestic, industrial and agricultural use
- generate hydroelectricity
- support a variety of recreational and leisure activities
- dispose of liquid waste products
- provide navigable waterways

The Colorado is a good example of a river basin with multiple uses:

- Lake Mead, impounded by the Hoover dam, provides water for Las Vegas.
- The Davis and Parker dams provide irrigation water for California's Imperial Valley and for central Arizona.
- The Hoover and Glen Canyon dams generate hydroelectricity for the cities of the southwest USA.
- Lake Powell and the Glen Canyon recreational area receive over 3 million visitors a year.

Examiner tip

Effective answers to questions on the multiple-resource use of rivers should refer to at least three or four different uses, supported by a range of actual examples.

Explain how rivers have influenced the growth of economic activities in one or more river basins.

The growth of economic activities depends on the availability of, and the demand for, resources — physical, economic and human.

Large navigable rivers like the Paraná in South America help to unlock the resource potential of interior Argentina, Brazil and Uruguay (as well as landlocked Paraguay and Bolivia) and stimulate demand. The development of grain growing and cattle ranching in the Pampas of Argentina was influenced by river ports such as Rosario and Santa Fé that enable the shipment of farm products to Europe and North America. The Paranà River is even more crucial to export-orientated farming in landlocked Bolivia and Paraguay.

Examiner tip

The best way to approach questions of this type is to recall the economic detail of river basins you have studied, and then organise your answer in terms of economic activities, e.g. transport, industry, energy generation, recreation and leisure.

In the arid and semi-arid Colorado Basin, water delivered by canal and pipelines from the Colorado River supports intensive irrigation agriculture in central Arizona and California's Imperial Valley. Water from the Colorado also supports the fast-growing cities of the southwest USA, including Las Vegas, with its huge tourism industry, the urban agglomeration of Phoenix-Scottsdale and Tucson. The growth of water-based recreation in the deserts of the southwest USA is due entirely to the damming of the Colorado and the creation of huge reservoirs such as Lake Mead and Lake Powell.

Explain how different land uses give rise to conflict in one or more river basins.

Land use in river basins includes urban development, farming, forestry, water catchment, recreation and leisure, conservation and energy production. The main conflicts in the Ribble drainage basin, in northwest England, arise when resource use by one activity degrades resources for another. Some activities (e.g. sewage treatment, industry, farming) lower water quality. Other activities (e.g. public water supply, irrigation for farming) divert water from the river and its tributaries or create artificial barriers such as weirs (e.g. industry). These activities conflict with recreational uses such as angling and canoeing, and with amenity and wildlife conservation by creating barriers to fish migration and lower water levels. Farming also conflicts with amenity and wildlife conservation through pollution caused by the runoff of chemical fertilisers (e.g. nitrates, reduced oxygen levels) and the spillage of silage liquor into streams and rivers.

Knowledge check 3

Make a list of, and explain briefly, possible conflicts that might arise between different land users in river basins.

What issues can arise from the development of river basins?

Key ideas	Content
The pressure to develop river basins can make them increasingly vulnerable to flooding	The study of at least one river basin to illustrate: • why some river basins are naturally prone to flooding, e.g. exceptional rainfall events, steep slopes, geology • how development can increase the flood risk, e.g. building on floodplains, land-use change including deforestation and urbanisation • the social, economic and environmental impacts of flooding, e.g. loss of life, disease, damage to infrastructure etc.

Key questions

Why are some rivers more naturally prone to flood than others?

Rivers flood when their discharge exceeds bankfull capacity. On average this can happen once or twice a year. Flooding is influenced by the speed of runoff and the amount of rain falling in a given time. Some rivers are naturally more prone to flooding than others due to factors such as relief, geology, **drainage density**, land use and climate.

The first four of these factors affect the speed of **runoff**. Upland catchments have steep slopes and faster runoff than lowland catchments. Runoff is more rapid on impermeable rocks (e.g. clay), which provide little natural storage, than on permeable rocks. (Note that natural lakes act as storage reservoirs and reduce the risk of flooding downstream.) Rainfall is transferred quickly to streams and rivers in catchments with dense drainage networks. Well-vegetated catchments (particularly afforested catchments) intercept large amounts of rainfall; this reduces the proportion of rainfall that becomes runoff and slows the movement of water to streams and rivers. Finally, flooding is more likely in catchments that experience extreme rainfall events. For example, **flash floods** are often generated in upland regions (e.g. Boscastle floods in 2004), whereas catchments in climatic zones that experience wet and dry seasons (e.g. monsoon, savanna) have regular seasonal floods.

Show how human activity and development within river basins can increase the risk of flooding.

Human activity can increase the risk of flooding by developing floodplains and changing land use in ways that accelerate runoff. Floodplains attract development because they offer flat, fertile land and provide locations for river crossings and transport routes. UK towns and cities at high risk from flooding (e.g. York, Chester, Carlisle, Shrewsbury, Tewkesbury) occupy floodplains. Despite the risks, there is still considerable pressure to develop floodplains. Much of the proposed Thames Gateway development alongside the Thames estuary is planned in locations that are flood prone.

Some changes of land use also increase flood risks. **Deforestation** accelerates runoff and increases river flow. Flooding in Bangladesh has increased in the past 20 or 30 years due to deforestation in the Himalayas. **Urbanisation** has a similar effect. Towns and cities cover natural permeable surfaces with impermeable concrete, tarmac and brick. Built-up areas have efficient artificial drainage systems that quickly transfer water to adjacent streams and rivers (see Figure 2). Changes in farming can also increase the flood risk. For example, replacing permanent pasture land with arable crops reduces interception and increases runoff.

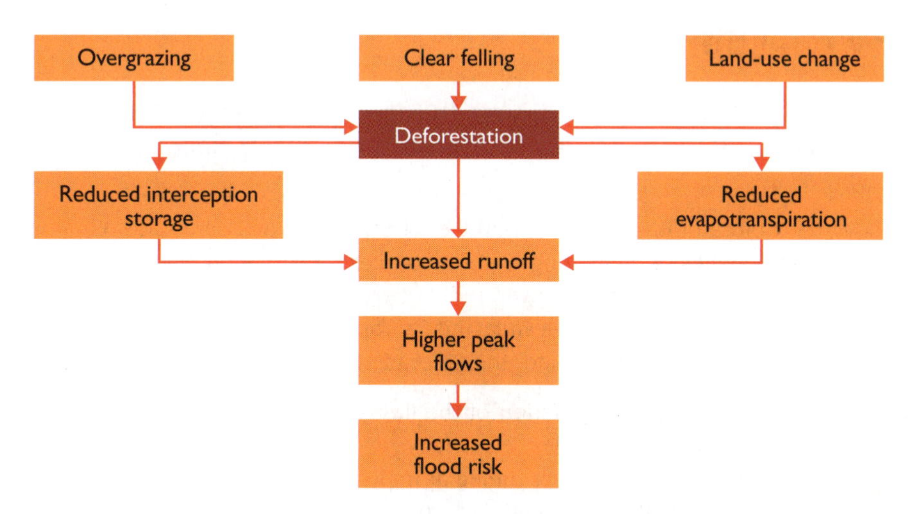

Figure 2 The effect of deforestation on rainfall–runoff relationships

Examiner tip

You should know that some of the physical factors that cause floods (e.g. drainage basin size and shape, relief, geology, soils) are fixed, while others (e.g. rainfall amount/type/intensity, evapotranspiration, vegetation cover, soil saturation) vary according to time of year. As a result, flood risks are often highly variable from season to season.

Examiner tip

Any investigation you undertake into specific flood events should include an assessment of the relative contribution of physical and human factors to flooding.

Knowledge check 4

List the main physical and human reasons for river floods.

With reference to one or more flood events, outline the social, economic and environmental consequences of flooding.

Prague, the capital city of the Czech Republic, was devastated by floods in August 2002. The cost of flood damage in the country amounted to €3 billion, and 110 people died.

As the flood wave on the River Vltava approached Prague, 50,000 people were evacuated from the city. Half of the evacuees were elderly and vulnerable. Sanitation and refuse collection services broke down, and power to hospitals was cut off. All bridges across the Vltava in Prague were closed to traffic. Many historic buildings in the Old Town were flooded, including the National Library, which housed important archives and historic collections. Prague's underground metro system was overwhelmed and it took 6 months and €230 million to repair the damage. Over 1,000 residential buildings in the inner city were flooded. Blocks of flats collapsed and other buildings became too dangerous to reoccupy. The worst affected districts had no gas or electricity for several days. Only one-third of the evacuees had returned to their homes by mid-November 2002.

Knowledge check 5

Outline the advantages and disadvantages of soft engineering flood management strategies.

What are the management challenges associated with the development of river landscapes?

Key ideas	Content
Successful management requires an understanding of physical processes	The study of at least two contrasting river basins to illustrate the varying need for planning and management in resolving development and flood risk issues, and possible land use conflicts in river basins.
Managing river landscapes is often about balancing socioeconomic and environmental needs — this requires detailed planning and management	

Key question

With reference to contrasting river basins explain how planning and management are helping to resolve problems of flooding and development.

Most of the Colorado Basin in the southwest USA is arid and semi-arid. Economic development depends on managing water resources for irrigation and domestic use. A series of dams (including the Hoover and Parker dams) in the lower basin provides water and electricity for Arizona, Nevada and California. Water resources are allocated to states in the Colorado Basin under the Colorado River Compact of 1922.

On the Paranà River, the construction of the Itaipu dam — a joint venture between Brazil and Paraguay — provides electricity to power economic development. Investments worth $475 million are aimed at improving navigation on the Paranà to help to unlock the agricultural potential of central Argentina.

In the UK, River Basin Management Plans encourage an integrated approach to river management. The first plan is for the River Ribble in northwest England. The plan will reduce the impact of flooding, promote sustainable water use and protect aquatic environments. Flood control is already well established: flood embankments protect urban areas in the lower basin; in the upper basin around Settle, a wide floodplain enables the temporary storage of floodwaters; and strict controls are in place to prevent building on the floodplain.

Examiner tip

You should adopt a critical view of river management strategies, which includes their environmental and socio-economic impacts as well as their sustainability.

Summary

- Fluvial landforms owe their formation to the interaction of sub-aerial and fluvial processes. Sub-aerial processes such as weathering and mass movement input sediment to streams and rivers. Fluvial erosion, transport and deposition are responsible for features such as river valleys, meanders and floodplains.
- Rivers are important resources and provide a range of economic and environmental opportunities. Rivers provide: water for irrigation, public water supply and cooling; transport arteries; opportunities for recreation and leisure activities; hydroelectric power; and conservation areas for wildlife.
- The development of river basins can increase the frequency and magnitude of flood events. Increased flood risks are often related to deforestation, urbanisation and human encroachment on floodplains.
- Flooding and land use conflicts associated with the development of river basins can be resolved by careful management and planning. However, effective strategies must strike a balance between socio-economic and environmental needs.

Coastal environments

What processes and factors are responsible for distinctive coastal landforms?

Key ideas	Content
Weathering, erosion, transport and deposition give rise to distinctive coastal landforms. These processes are influenced by a range of factors, which vary from place to place.	The study of an extended stretch of coastline or coastlines to illustrate: • a range of coastal erosional features, e.g. headlands, cliffs, arches, stacks, shore (wave-cut) platforms • a range of coastal depositional features, e.g. beach types such as spits, tombolos, barrier beaches, dunes • the processes responsible for these erosional and depositional features, including wave action, e.g. marine processes, sub-aerial processes (weathering, mass movements), aeolian processes (i.e. sand transport by the wind) • the factors affecting the development of coastal features including rock type and structure, aspect and sea-level change

Key questions

How do weathering and slope processes influence the development of coastal landforms?

Weathering and slope processes are active on cliffs on upland coasts. Exposed rocks are attacked by **physical** and **chemical weathering**. These processes include:

- **freeze–thaw** (especially on well-jointed rocks)
- corrosion on carbonate rocks such as limestone
- hydrolysis on granite
- wetting and drying (hydration)

Examiner tip

It is useful to consider how sub-aerial processes in coastal environments differ from those at inland sites. On the coast, wetting and drying and salt spray promote hydration, solution and salt weathering, while steep slopes and undercutting of cliffs by wave action trigger mass movements.

Weathering contributes to the retreat of cliffs and to **rockfall**. Slope processes such as **soil creep**, **rotational slumping**, **mudflows** and **mudslides** are most effective on weaker, less coherent rocks such as clay and shale.

Mass movement is particularly active on coasts such as Holderness (in Yorkshire) and west Dorset. At Holderness, slumping is responsible for the steep profile of boulder clay cliffs. At Black Ven in Dorset, giant mudslides in Lias clay give a step-like profile to the coastal slope. In many parts of northern Britain, soft boulder clay rests above more resistant rocks such as sandstone and chalk. Because mass movement processes are more effective on clay, the clay often forms a shallow slope of 40–45°, with the lower section of sandstone or chalk being almost vertical.

Knowledge check 6

What is the difference between weathering and mass movement?

How do rock type, structure and aspect influence the development of coastal features?

Resistant rocks often produce cliffs, caves, arches and stacks on upland coasts. All the major stretches of upland coast in the British Isles comprise resistant rocks (e.g. Torridon sandstone in northwest Scotland, Jurassic sandstone in northeast Yorkshire, chalk and basalt in County Antrim). These rocks maintain steep slopes and, because they crop out on upland coasts, create high and almost vertical cliffs. Despite their resistance wave action exploits lines of weakness (**joints**, **faults**, **bedding planes**) in these hard rocks, leading to a succession of erosional features — first caves, then arches, stacks and stumps, and finally shore (or wave-cut) platforms. Less resistant rocks generally form lowland coastlines, dominated by depositional features such as beaches, dunes, salt marshes and mudflats (e.g. north Norfolk).

Structure affects the **planform** of a coastline (i.e. the shape of the coastline seen on a map). The classic example is the coast of Dorset. There several sedimentary rock types of variable resistance run parallel to the south coast and at right angles to the east coast (see Figure 3). As a result the south coast, fronted by a single rock type of uniform resistance (i.e. limestone), is comparatively straight. This is known as a **concordant** coastline. In contrast, on the east coast where all the different rock types crop out, the weaker clays, sands and gravels have been eroded into broad bays (i.e. Swanage Bay, Studland Bay). Meanwhile the resistant limestone and chalk form prominent headlands such as Peveril Point and The Foreland.

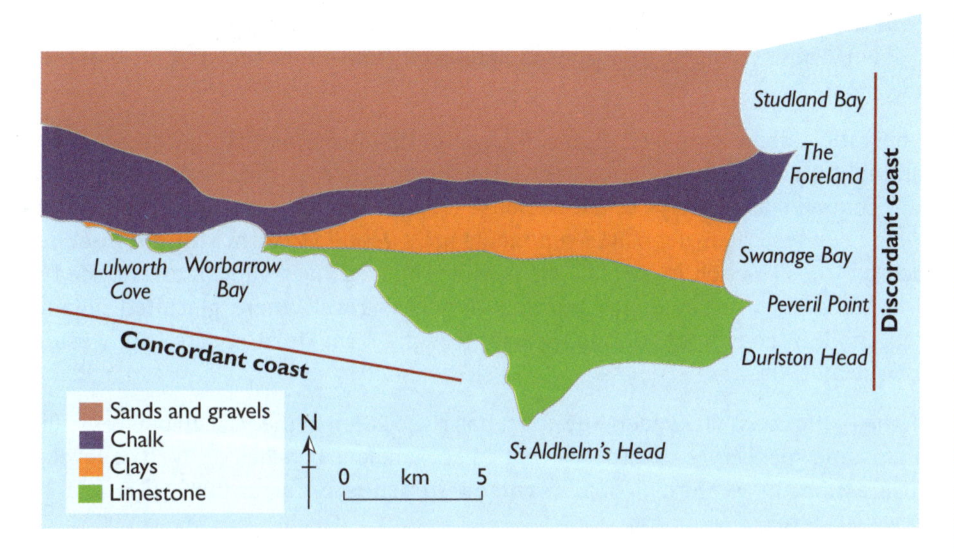

Figure 3 Rock structure and coastal planform, Isle of Purbeck, Dorset

Aspect is the direction a coastline faces. Coastlines exposed to large expanses of open ocean have long **fetches** and experience high-energy waves. Examples of such coastlines are those in southern Chile and in Cape Town, South Africa. Where resistant rocks meet high-energy coastlines, spectacular erosional features often develop. Low-energy coastlines occur where the fetch is short. In the Leeward Islands in the West Indies, the Caribbean coasts have a short fetch and low energy. In contrast to the Atlantic coasts where wave energy is high, extensive beaches and coral reefs form.

How do spatial differences in wave energy affect the shape of the coastline?

Spatial differences in wave energy on coastlines are due to differences in aspect (see previous key question) and **wave refraction**. As waves approach a coastline they bend or refract. Refraction is due to the slowing of waves as they enter shallow water in the inshore zone. The impact of aspect and refraction is to create unequal distributions of wave energy along coastlines. Thus coasts with a long fetch are high-energy environments while sheltered coasts experience low energy. Wave refraction concentrates energy on headlands, but disperses it in bays.

Where wave energy is concentrated (assuming that other variables such as rock type, depth of sea bed, relief etc. are constant) erosional features such as cliffs, caves, arches, stacks and shore platforms develop. In low energy environments, depositional features such as beaches, mudflats and salt marshes form.

How does sea level change modify coastal landforms?

A worldwide rise or fall in sea level is known as a **eustatic** change. This is an absolute change, caused by either an increase or decrease in the volume of water in the oceans. It is associated with **glacial** (falling sea level) and **interglacial** (rising

Examiner tip

Waves are often labelled as either destructive or constructive. These terms refer to the effectiveness of waves as agents of sediment transport. Thus, destructive waves have high energy. They erode sediment from beaches and deposit it offshore. Constructive waves are lower-energy waves, depositing sediment and building beaches.

sea level) periods. Relative changes in sea level occur when land masses shift up or down. These changes are localised and result from either **tectonic** or **isostatic** movement.

A eustatic fall in sea level (typically up to 100m) exposes large areas of the continental shelf. Rivers deposit fluvial sediments across these areas and incise their channels and valleys to a lower base level. When the climate warms and sea levels rise these river-deposited sediments are swept up to form **shingle beaches** such as Chesil Beach in Dorset. Meanwhile, incised river valleys are flooded to form steep-sided, branching estuaries known as **rias**. Where glaciated uplands meet modern coastlines, flooded glacial troughs form **fjord** coasts (e.g. western Norway and Alaska).

Knowledge check 7

Describe two ways in which rising sea level has affected coastal landforms.

On the Baltic coast of Sweden, the coast has risen 285m in the past 10,000 years due to isostatic recovery — a process that raised ancient beaches above the level of wave action. In western Scotland, **raised beaches** occur at 30m, 15m and 8m above sea level.

How and why do beaches vary in plan and profile?

Beach plans describe the shape of beaches as seen on a map. **Beach profiles** refer to the cross-sectional shape beaches from the high-water mark to the low-water mark. In planform, beaches may be straight or curved. They can be joined to the coast along their entire length, joined to the coast at one end, or detached from the coast. Straight beaches, attached to the coast in their entirety, form where waves are fully refracted and break parallel to the shoreline. They are known as **swash-aligned** beaches.

Beaches in bays and coves are usually swash-aligned. Straight beaches such as Chesil Beach in Dorset and Slapton Sands in Devon are also swash-aligned. They formed between 6,000 and 18,000 years ago when bars of shingle migrated onshore as sea level rose. When sea levels eventually stabilised, it was purely by chance that Chesil linked the coast of Dorset to the Isle of Portland. Chesil Beach, connecting the mainland to an island, is therefore classed as a **tombolo**. Slapton Sands has a similar origin but is known as a **barrier beach**. Its final location was across a bay, where it created a brackish lagoon (Slapton Ley) on its landward side.

Examiner tip

Textbooks classify beach types according to their shape or planform rather than the processes responsible for their formation (i.e. swash- or drift-aligned). Thus beaches such as spits, tombolos and barrier beaches could be either swash- or drift-aligned features.

Where waves break obliquely to the shoreline, **longshore drift** moves sediment along the coast. The resulting beach forms are **drift-aligned**. The best-known drift-aligned beaches are spits: they grow outwards from the coast (often across estuaries) and are attached to the coast at one end (the **proximal** end). Typically, their unattached or **distal** end is hooked or **recurved**.

Often a sequence of old recurves can be seen along the length of a spit (e.g. Blakeney Point in north Norfolk). They show stages in its growth and provide evidence that growth was due to longshore drift. The distal ends of spits often support small dunes; the landward side is an area of low energy, shallow water, mudflats and salt marshes.

How can coasts be protected from the effects of natural processes?

Key ideas	Content
There are a number of ways that coastal areas can be protected, ranging from hard engineering and managed retreat	The study of an extended stretch of coastline (e.g. the Yorkshire coast, Christchurch Bay) to illustrate: • the reasons why some coastal areas need protecting • the different methods of coastal protection, including hard and soft engineering and managed retreat • the planning, management and environmental issues associated with different methods of coastal protection

Key questions

Why do some coastal areas need protection from natural processes?

Some coastal areas need protection from the natural processes of erosion and flooding because the cost of losing property and infrastructure is economically and politically unacceptable. Major coastal settlements that are vulnerable to erosion or flooding in the UK are protected by **hard engineering** structures maintained by the Environment Agency, Defra and local authorities. On the Yorkshire coast between Flamborough and Kettleness, hard defences (sea walls and armour blocks) protect the coastal towns of Whitby, Scarborough and Filey against erosion and flooding.

Coastal protection is especially urgent where rates of erosion are rapid and threaten urban areas. Christchurch Bay in Hampshire has one of the most rapidly eroding coastlines in the UK. Rapid erosion is due to:

- unstable sands and gravels resting on impermeable clay
- dominant high energy waves from the southwest, where the fetch is longest
- narrow beaches that are unable to absorb wave energy

Examiner tip

There are two quite separate protection issues that affect human populations in coastal areas: erosion and flooding. They should not be confused: even though their causes overlap, the human responses are often very different.

What methods are available to protect coastlines from natural processes?

The two methods available to protect coastlines from natural processes are hard engineering and soft engineering. Hard engineering involves building structures to stop erosion and flooding. These structures include seawalls, gabions, revetments, groynes, breakwaters, embankments and armour blocks. They protect most large settlements around the UK coastline (see Figure 4 on p. 18). Hard engineering structures are costly to build and maintain, and in the long-term are often **unsustainable**.

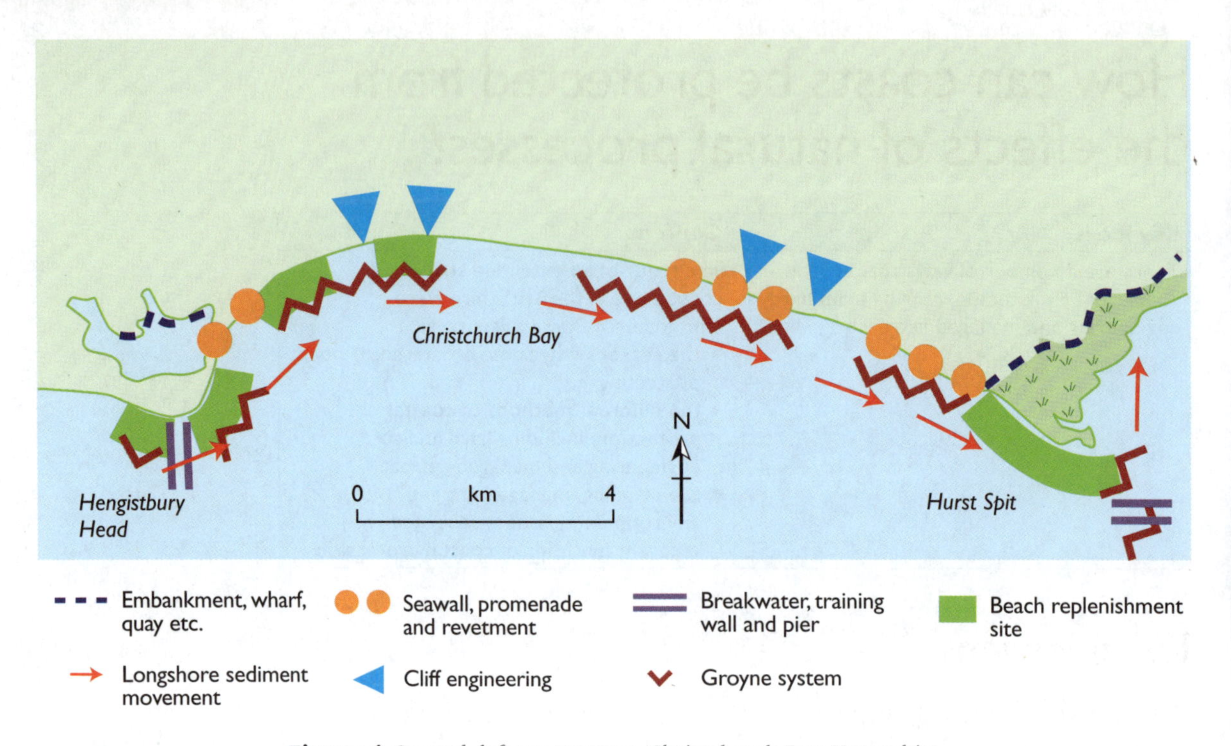

Figure 4 Coastal defence systems, Christchurch Bay, Hampshire

Soft engineering, such as beach replenishment, is less intrusive environmentally. **Beach replenishment** involves importing sand and shingle to strengthen natural beaches. A broad beach is an effective way of dissipating wave energy. **Managed realignment**, where existing hard defences are either abandoned or set back, is another soft engineering response. The sea is allowed to flood areas to the landward side of the old defences; it gradually forms mudflats and salt marshes. Mudflats and salt marshes, like beaches, absorb wave energy, making the new defence lines sustainable.

What planning, management and environmental issues are associated with the different methods of coastal protection?

Hard engineering structures raise a number of issues. Most hard structures are expensive to build and can easily cost more than the value of the properties they defend. Seawalls are most expensive (typically £10 million per kilometre). In addition, seawalls and other hard defences require constant maintenance — and with rising sea levels and increasingly stormy conditions possible in the future (due to climate change), the cost of maintenance is likely to rise steeply.

The issue is whether all hard defences should be maintained, regardless of cost. Where hard defences are abandoned and nature is left to take its course (e.g. Happisburgh in Norfolk) problems arise when homes and infrastructures are threatened. Although in the long term the coastline will eventually establish a new equilibrium, in the short term there will be considerable disruption and economic and social costs.

Most hard engineering structures are ugly and environmentally intrusive. This can be an issue, especially along coastlines of high amenity value. In addition, the structures often limit public access to beaches and the shoreline.

Hard engineering structures also interfere with the **coastal sediment budget**. Sea walls stop erosion and halt the input of fresh sediment to the coastal system that would otherwise build natural defences such as beaches, mudflats and salt marshes. Sea walls also create problems of **coastal 'squeeze'** and the erosion of salt marshes and mudflats. Groynes can be problematic if they prevent the longshore movement of beach sediment. In these circumstances beaches downdrift are starved of sand and shingle, accelerating erosion in those areas.

The policy of managed realignment is equally controversial. By setting back coastal defences, large areas of farmland (often high quality arable land in eastern England) are lost to the sea. Although this brings benefits for wildlife, there is a considerable economic cost in lost farm production.

Examiner tip

You will need to know the advantages and disadvantages of hard and soft engineering approaches to coastal erosion and flooding. The most important issue to address is the economic and environmental sustainability of these strategies.

Knowledge check 8

What are the main differences between hard and soft engineering approaches to coastal management?

In what ways can coastal areas be a valuable economic and environmental resource?

Key ideas	Content
Coastal areas provide opportunities for a number of human activities, including: • industrial development • transportation • residential development • energy development • recreation and leisure • conservation	The study of at least two contrasting coastal environments (e.g. southern Spain, Bangladesh) to illustrate: • the variety of activities found in coastal areas • the reasons for the growth and development of these activities • the conflicts that may result from the growth and development of these activities

Key questions

What are the main human activities in coastal areas and why have they developed?

Economic activities in coastal areas include fishing, trade (ports), industry, tourism and conservation. The distribution of these activities reflects the availability of resources and the advantages of location in the coastal zone.

In Bangladesh, the coastal ecosystems are rich in nutrients and support a valuable fishing industry, including shrimp farming. Major industrial concentrations around Chittagong and Khulna are based on the manufacture of jute, textiles, pulp and paper,

Examiner tip

Effective answers to questions dealing with human activities found in coastal areas must consider a range — at least three or four — of different activities, supported with appropriate examples and/or case studies.

shipbreaking and so on. The coast provides suitable sites for industry and access to sea and river transport for exports and the procurement of materials.

Tourism is the main activity on Spain's Costa del Sol. Its growth and development are explained by its hot summer climate, warm sea, beaches and easy access (by budget airlines) for holidaymakers in the UK, Germany, the Low Countries and Scandinavia.

What conflicts arise from the growth and development of human activities in coastal areas?

Conflicts in coastal zones often arise between economic development and the environment.

Knowledge check 9

List the potential conflicts between economic development and the physical environment in the coastal zone.

In Bangladesh, industrial growth has polluted coastal waters with untreated sewage, agrochemicals (pesticides) and heavy metals. Economic development (especially shrimp farming) has led to deforestation of coastal mangrove forests. These forests provided vital timber and fuelwood resources for local people. The surviving forests still give protection against tidal surges and support a rich and biodiverse ecosystem, but are under threat.

Conflicts on the Costa del Sol include rapid and unplanned urbanisation that is destroying the natural environment. Tourism and golf courses have put pressure on limited water resources, while inadequate sewage treatment has produced low quality bathing water in many places.

What are the management challenges associated with the development of coastlines?

Key ideas	Content
Successful management requires an understanding of physical processes	The study of at least two contrasting coastal areas to illustrate the varying need for planning and management in resolving development issues and conflicts in such areas
Managing coastal areas is often about socioeconomic and environmental needs — this requires detailed planning and management	

Key question

With reference to contrasting coastal areas explain how planning and management can help to resolve problems of development and conflict.

In Bangladesh's coastal zone, an integrated management approach to problems such as natural disasters, sea defences, salt water intrusion and the conservation of ecosystems has been used since 1999. Most of these problems are interrelated, and integrated management by a single agency offers the best chance of finding solutions.

In England and Wales, **Shoreline Management Plans** (SMPs) are the main planning instruments. The coastline at Christchurch Bay is heavily managed to address problems such as cliff erosion at Barton and Highcliffe, the reduced flow of longshore sediment to Hurst Spit, and the erosion of saltmarsh in the lee of Hurst Spit. Current strategy includes:

- managed retreat in the west, allowing the cliffs to erode naturally and input sediment into the coastal system
- at Barton, Christchurch and Highcliffe, existing sea defences (seawalls, groynes and revetments) are being maintained and in places the cliffs will be regraded and drained
- no active intervention between Barton, Milford-on-Sea and the eastern end of Hurst Spit

Soft engineering, through beach replenishment with sand mined offshore, will compensate for the loss of beach sediment at Hurst Spit, Mudeford Spit and Christchurch.

On the Costa del Sol, where thousands of homes and some hotels have been built illegally (i.e. without planning permission), coastal authorities have adopted strong but controversial policies, which include demolition. Modern planning aims to limit future development and tackle the problems of traffic congestion, water shortages and environmental degradation. Many of the worst examples of cheap, poorly planned development from the 1960s and 1970s have already been removed.

Examiner tip

Coastal management policies may allow the loss of land and property to erosion and flooding. These policies are controversial. Case studies, with arguments for and against management policies, need to be learned if issues are to be addressed convincingly.

Summary

- The interaction of sub-aerial and marine processes gives rise to distinctive landforms in coastal environments. Sub-aerial processes include weathering and mass movement; marine processes are wave erosion, transport and deposition; and the transport and deposition of sediment by tidal currents. Along dune coastlines, wind (aeolian processes) also has an important influence on coastal landforms.
- The processes responsible for coastal landforms vary from place to place, depending on local factors such as rock type, geological structure, relief and wave energy.
- Sea level change, associated with glacial and inter-glacial periods, produces a suite of coastal landforms, including fjords, rias, estuaries, shingle beaches and raised beaches.
- Coastal protection operates where coastal erosion and flooding are significant threats to life and property. There are two approaches to coastal protection: hard engineering (e.g. seawalls, groynes) and soft engineering (e.g. beach replenishment, managed realignment).
- Coastal environments provide valuable resources for economic activities such as trade, industry and tourism, and habitats for wildlife. The development of coastlines often leads to conflicts between economic activities and between economic activities and conservation.
- Management of the coast in the UK is influenced by the economic and environmental costs and benefits of particular strategies and their sustainability. Shoreline Management Plans, based on an understanding of coastal environmental systems, inform and guide debates on development issues.

Cold environments

What processes and factors give cold environments their distinctive characteristics?

Key ideas	Content
The distinctive characteristics of cold environments result from climatic and geomorphological processes	The study of cold environments to illustrate: • the impact of climate and weathering on the physical landscape • the way that ice and water shape the landscape to produce distinctive landforms, including cirques, arêtes, U-shaped valleys, waterfalls, lakes, moraines and outwash plains
These processes are influenced by a range of factors that vary from place to place	

Key questions

How do climate and weathering influence the physical landscape in cold environments?

Glaciers and ice sheets develop in cold, humid climates, where rates of **accumulation** of snow and ice exceed rates of **ablation** (i.e. melting, sublimation). Depending on the severity of the climate and local topography, glaciation may result in ice sheets, which submerge the landscape, or valley glaciers, originating in cirques or icefields. Glaciation greatly modifies pre-existing landscapes, creating new erosional and depositional features (e.g. U-shaped valleys, cirques, moraines and glacio-fluvial landforms).

In warmer glacial climates, meltwater allows glaciers (**warm-based**) to move by sliding, greatly increasing their erosive powers. In cold glacial climates, with an absence of meltwater, glaciers are often frozen to the ground surface. Glacier movement is by **internal deformation** with only limited erosion. Thus, **cold-based** glaciers protect ice-covered landscapes from erosion.

Examiner tip

Explanations of freeze–thaw weathering must make it clear that it is the frequency of freeze–thaw cycles (i.e. the number of days when temperatures fluctuate above and below freezing) and not their severity that determines the effectiveness of frost action.

Weathering is strongly controlled by climate. The dominant weathering process in cold climates is **frost weathering**, by freeze–thaw. The effectiveness of freeze–thaw depends on the number of **freeze–thaw cycles**, when temperatures fluctuate above and below freezing. Periglacial and glaciated mountain climates experience frequent freeze–thaw cycles, resulting in features such as **talus slopes** and **blockfields**.

In glaciated regions, the ground surface may be broken up by freeze–thaw prior to the advance of glaciers, greatly increasing the effectiveness of erosion by glaciers and ice sheets. Areas at altitude not submerged by ice were at various times exposed to severe frost action. On steep slopes this produced extensive **screes**; on plateau-like surfaces **blockfields** and **tors** developed.

Frost weathering also inputs weathered rock debris to glaciers from valley slopes (by mass movements such as rock avalanches), providing glaciers with the 'tools' for **abrasion** and contributing to the formation of ice-contact depositional features such as **lateral moraines**, **terminal moraines** and **lodgement tills**.

How do ice and water shape the landscape to produce distinctive landforms in cold environments?

Valley glaciers erode upland landscapes, through **abrasion** and **plucking**, to form classic features such as **cirques**, **arêtes**, **glacial troughs** and **hanging valleys**. Continental ice sheets create erosional features such as **roches moutonnées** and **knock-and-lochan topography**. Abrasion is the result of the scouring of rock surfaces by rock fragments embedded in the ice. Plucking occurs when pressure at the base of a glacier causes localised melting. Meltwater penetrates rock joints, re-freezes and removes (or plucks/quarries) rock fragments. Frost action through freeze–thaw also plays a part in shaping the landscape in areas not covered by ice (e.g. ridges, high valley slopes). Glaciers and ice sheets transport rock debris or moraine deposited by moving and/or stagnating ice. The resulting ice-contact depositional features include **moraines** (terminal, recessional, push, lateral), **drumlins** and **till plains**.

During deglaciation, large amounts of meltwater flow on, within and beneath glaciers and ice sheets. Meltwater lakes, formed by temporary ice dams, can overflow to erode **meltwater channels**. More often, meltwater streams carrying huge sediment loads form depositional features such as **eskers**, **kames**, **kame terraces** and **outwash plains**. Ice and water also help to shape the landscape in periglacial regions.

Knowledge check 10

What is the difference between glacial moraines and glacio-fluvial deposits?

Examiner tip

The relationship between glaciation and landscape modification is complex. Good exam answers will show an awareness of the complexity — an understanding that there have been many glacial episodes in the past 2 million years; and that there are different types of glaciation (e.g. continental — Antarctica, valley — Alps).

Why are cold environments considered to be fragile?

Key ideas	Content
Climatic extremes lead to finely balanced ecosystems that can be easily damaged	The study of one cold environment (e.g. tundra) to illustrate: • the impact of climate on the nature of the ecosystem • how physical and human factors make the environment ecologically vulnerable
Both flora and fauna can suffer as a result of change, and regeneration is difficult in the harsh conditions	

Key questions

With reference to a specific cold environment, explain how climate influences the nature of the ecosystem.

Tundra ecosystems are found in arctic (high latitude) and alpine (high mountain) environments. Both are harsh environments for plants and animals. Climatic

problems include low temperatures, prolonged snow cover, strong winds and a short growing season. Ecosystems are dominated by low-growing, **perennial** plants. With the exception of dwarf birch and willow, trees are absent. Typical plant species are mosses, sedges, lichens and woody shrubs such as heather, ling, bilberry and mountain avens. A few herbaceous flowering plants have also adapted to the tundra environment (e.g. moss campion, snow gentian). Severe climatic conditions mean that growth, **net primary productivity** and **biodiversity** are low.

Specific plant adaptations to climate include:

- dark leaves to absorb insolation; cushions and rosettes; and flowers that track the sun — all designed to create microclimates that raise temperatures
- short life cycles and the storage of food in tubers and rhizomes allow plants to tolerate a snow cover that may last 10 months a year
- low growth habit for shelter against strong winds
- perennial habit to combat short growing seasons

Animal populations are sparse because of low plant productivity. **Food webs** are short (see Figure 5). Larger mammals (e.g. caribou) and birds migrate to the arctic tundra in summer to feed on the seasonal glut of insects, herbs and berries. Because weather conditions in any single year are unpredictable, numbers of rodents, raptors and carnivores (e.g. foxes, stoats) follow 'boom and bust' cycles.

Examiner tip

A useful approach to the analysis of ecosystems is to think of their main features: biodiversity, productivity, stability, nutrient cycling and the length or complexity of food webs and food chains.

Knowledge check 11

State three ways in which climate influences tundra ecosystems.

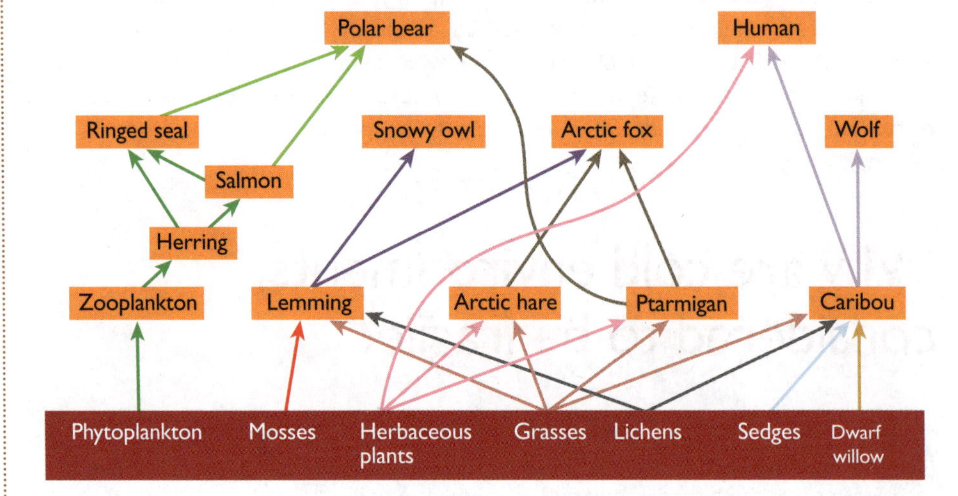

Figure 5 Arctic tundra food web

With reference to a specific cold environment, explain how physical and human factors make the environment ecologically vulnerable.

The arctic tundra in northern Alaska (the North Slope) is an ecologically fragile environment that is easily degraded by human activity. It is a vast area of low relief underlain by **permafrost**. The climate is extreme: winters are long, dry and very cold; summers are brief, moist and cool (see Figure 6). Day length varies from continuous darkness in winter to 24 hours' daylight in summer.

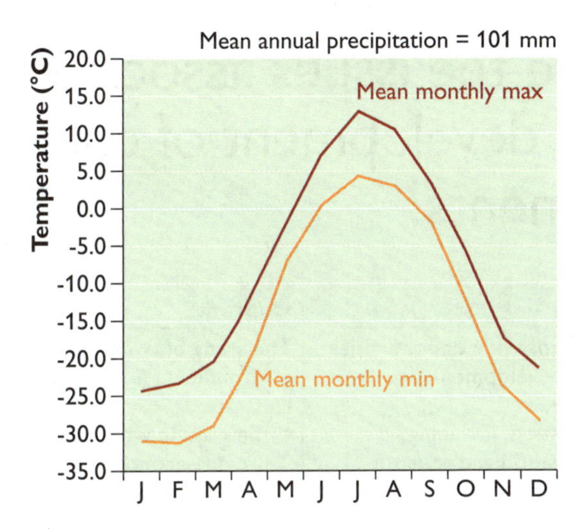

Figure 6 Mean monthly temperatures at Prudhoe Bay on Alaska's North Slope coast

Because of the harsh environment, **primary productivity** and biodiversity are low. Low-growing vegetation, adapted to the exposed conditions, comprises grasses, sedges, lichens, mosses, heathers and dwarf birch. Soils are acidic; frozen in winter and waterlogged in summer. Food webs are short and highly interconnected, and nutrient cycles operate on tight budgets.

Tundra ecosystems are fragile because:

- few species occupy each ecological niche within simple food webs
- animal populations are sparse
- rates of soil formation and plant growth are slow and therefore recovery from any environmental change is also slow
- disruption to vegetation cover leads to melting of the permafrost, which triggers major ecological change

Human activity has the potential to cause widespread degradation in the arctic tundra. Indigenous people have lived in balance with the tundra ecosystem for thousands of years, but recent industrial development on the North Slope, based on oil and gas extraction, has adversely impacted the environment.

- The Trans-Alaskan pipeline system has disrupted the seasonal migration of caribou herds.
- Oil spillages have degraded local habitats.
- Seismic exploration has disturbed wild animal populations.
- Gravel extraction from river beds has damaged fish populations.
- Landfill and waste tips around oil and gas installations have increased populations of predators and scavengers such as foxes and ravens.
- Frozen ground and slow rates of organic decomposition create problems for solid waste disposal.

Knowledge check 12

Give reasons why development in the tundra often degrades local ecosystems.

What are the issues associated with the development of cold environments?

Key ideas	Content
Cold environments provide opportunities and challenges for development. Opportunities include: • resource exploitation, including agriculture, recreation and tourism Challenges include: • environmental constraints • costs/remoteness • conflicts with indigenous populations	The study of two contrasting cold environments (e.g. Alaska, the Himalayas) to illustrate: • the ways in which cold environments provide economic opportunities such as resource exploitation and recreation and tourism • the ways in which the development of cold environments presents social, economic and environmental challenges, including: conflicts with indigenous populations; costs of development; environmental impacts

Key questions

In what ways do cold environments provide economic opportunities for development?

Cold environments provide economic opportunities for development through recreation and tourism, and resource exploitation. Mountain environments such as the Alps and the Rockies support winter sports (skiing), as well as providing opportunities for summer recreational activities such as walking, climbing and canoeing. Tourism has expanded rapidly in the Himalayas in Nepal. The Annapurna region receives 60,000 trekkers a year. Visitors are attracted by the spectacular mountain scenery and extraordinary biodiversity.

Examiner tip

Recall of opportunities for development in any environment is best achieved by (a) identifying possible economic activities and (b) examining each relevant activity within the context of a specific example.

Large areas of tundra in Alaska, northern Canada, and Siberia have extensive oil and gas reserves. The North Slope of Alaska has seen large-scale exploitation of oil and gas since the 1960s. Prudhoe Bay is North America's largest oilfield. Overall, 4,300 oil and gas wells have been drilled in Alaska. Currently (and controversially) new oil and gas fields are scheduled for development in Alaska's Arctic National Wildlife Refuge (ANWR).

How do cold environments present social, economic and environmental challenges to development?

Cold environments present social and economic challenges to development because:

- Development often conflicts with the interests of indigenous people.
- Cold environments are often remote and isolated, which increases development costs.
- Extreme cold climates support fragile ecosystems that are easily damaged by human activity.

Oil and gas exploration and development in Alaska's ANWR has impacted on the native people, including the Gwich'in and Inupiaq Indian tribes. Both groups practise traditional subsistence economies. The Gwich'in follow the caribou herds and the Inupiaq hunt whales, seals and walruses. Oil and gas developments could disrupt the behaviour of the caribou and affect their main calving grounds. These industries can also threaten marine ecosystems by using seismic techniques to prospect for oil, and by oil spillages and the pollution of inshore waters.

Cold environments such as Alaska are sparsely populated with limited transport infrastructures. Development involves the construction of new roads, pipelines and settlements, which add significantly to costs. The permafrost must be insulated to protect it from thawing. This is crucial for successful economic development — yet costly.

Given the fragile nature of tundra and mountain ecosystems, the challenge is to achieve environmentally sustainable development. Huge numbers of visitors to the Himalayas have already degraded ecosystems in the most popular areas, and have contributed to deforestation, trail erosion and pollution (e.g. by creating human waste and large quantities of non-biodegradable junk).

Examiner tip

From a geographical perspective, development in extreme environments faces challenges of remoteness/lack of accessibility, inadequate infrastructure and unsustainable impacts on fragile ecosystems. Any assessment of the challenges posed must weigh the economic benefits against the social, economic and environmental costs.

How can cold environments be managed to ensure sustainability?

Key ideas	Content
Managing cold environments is often about balancing socioeconomic and environmental needs — this requires careful management to ensure sustainability	The study of two contrasting cold environments (e.g. Alaska, the Himalayas) to illustrate: • how such fragile environments can be exploited for short-term gains • how careful management can help to ensure sustainable developments in fragile environments

Key question

With reference to two contrasting examples, explain how careful management can help to ensure sustainable developments in cold environments.

Sustainable development in cold environments does not cause long-term damage to local ecosystems. Careful management of two cold environments — Alaska's North Slope and the Annapurna conservation area in Nepalese Himalaya — aims to make economic development sustainable.

Drilling for oil and gas in Alaska employs new technology designed to reduce its environmental footprint. Seismic exploration for oil and gas reduces the need for drilling and building access roads on the tundra. Where new roads are needed they are built on insulated ice pads that protect the permafrost from melting. Powerful computers help locate oil and gas-bearing structures, and new drilling techniques allow oil and gas deposits to be accessed several kilometres away from the drilling rigs. All this technology makes drilling environmentally less intrusive.

In Annapurna, the environment has been degraded by local farmers (through **overgrazing** and **deforestation**) and international tourism. The area's management plan is community-based and its success rests on the support of local people. **Reafforestation**, water and wildlife conservation programmes not only help the environment but also provide employment for local people — for example, in tree nurseries and as forest guards. Protecting the environment and its biodiversity is also good for tourism, which creates jobs for local people in hotels and visitor centres, and as trekking guides. Foreign tourists are charged US$7 to enter the Annapurna area. This money is used to help to maintain trails and improve sanitation facilities. Forests are conserved by developing alternative energy sources to fuelwood (e.g. hydroelectric power, kerosene stoves), and by family-planning schemes, which in the long-term will reduce pressure on environmental resources.

Knowledge check 13

Suggest ways in which development in cold environments can be achieved sustainably.

Summary

- Cold environments comprise high mountain regions and the sub-arctic tundra. Frost action and freeze–thaw dominate weathering in cold environments. In upland areas they create their own landforms, such as talus (scree) slopes and tors.
- Cold environments are either glaciated, or have been glaciated within the past 20,000 years. During glacial periods, glaciers, ice sheets and meltwater mould the landscapes into distinctive forms. In upland areas glaciers have sculpted classic erosional features such as cirques, arêtes and U-shaped valleys. In contrast, glacial and glacio-fluvial deposition in cold environments was concentrated in lowland areas. These areas support a variety of depositional landforms such as moraines, eskers and outwash plains.
- Development in cold environments centres on the extraction of energy resources (especially oil and gas) and the exploitation of tourism opportunities. Without management, development can have adverse effects on the landscape, ecosystems and indigenous people.
- Development in cold environments requires management to ensure its sustainability. This can be done by: introducing new technologies (e.g. in oil and gas extraction) to minimise the environmental impact; protecting the permafrost and its delicate thermal balance; and providing economic assistance to indigenous people to diversify employment and provide financial incentives to conserve environmental resources (e.g. forests).

Hot arid and semi-arid environments

What processes and factors give hot arid and semi-arid environments their distinctive characteristics?

Key ideas	Content
The distinctive characteristics of hot arid and semi-arid environments are a result of climatic and geomorphological processes	The study of hot/semi-arid environments to illustrate: • the impact of climate and weathering on the physical landscape • the way that wind and water shape the landscape to produce distinctive landforms, including sand dunes, canyons and canyon landscapes, sculptured rocks, wadis and salt pans
These processes are influenced by a range of factors that vary from place to place	

Key questions

How do climate and weathering influence the physical landscape in hot arid and semi-arid environments?

Climate has a major influence on types and rates of weathering in hot arid and semi-arid environments. Even in the most arid climates it rains occasionally; moisture is also available from dew and fog. **Hydration** — the absorption of water by minerals, which increase in volume and cause stress in rocks — is fairly widespread and leads to the flaking of rocks exposed near the surface (exfoliation).

Freeze–thaw can occur in hot arid and semi-arid environments during winter nights, especially at altitude (e.g. Arizona, Namibia) when temperatures fall below zero. **Salt weathering** occurs when salt in rocks crystallises out of solution. The growth of salt crystals stresses the rock causing flaking and rock disintegration. Salts such as sodium chloride and sodium sulphate, which in humid environments are removed by streams and rivers, accumulate in hot arid and semi-arid environments.

Cloudless skies combined with high surface temperatures during the day and low temperatures at night cause rock minerals to expand and contract. When this thermal process occurs in the presence of moisture (from rainfall or dew) it leads to **insolation weathering**. This is another process that contributes to the peeling of surface rock layers known as exfoliation.

Climate and weathering are responsible for rock breakdown, the parallel retreat of scarp/plateau slopes, and the formation of features such as natural arches and alcoves. They are also the source of much of the sand found in 'erg' environments.

Examiner tip

Because of their aridity it is easy to forget that water (in the form of atmospheric moisture) is present in deserts. This moisture, together with occasional rainfall, plays a crucial role in desert weathering. Also, remember that many tropical and sub-tropical deserts may be 1,000–2,000m above sea level and often experience frost in the winter.

Knowledge check 14

What are the differences between the following aeolian (wind) processes: abrasion, saltation and deflation?

Examiner tip

Because it rains infrequently in deserts, the influence of water in shaping desert landscapes is often overlooked. You need to know that water's effectiveness (as an agent of erosion) is due to factors such as vegetation cover, slopes, and the permeability of surfaces as well as rainfall amounts. Also remember that many desert landforms may be relict features, formed during wetter climatic periods.

How do wind and water shape the landscape in hot arid and semi-arid environments?

Wind and water are the two main agents of erosion in hot arid and semi-arid environments. Wind is an effective erosional agent in this type of environment because of the lack of vegetation cover and the dry soil and regolith. The main effect of the wind is the removal of fine, loose-grained particles in a process called **deflation**. The sand-blasting effect of particles **entrained** and **saltated** by the wind sculpts minor landforms such as **yardangs** — streamlined wind-eroded ridges aligned in the direction of the prevailing wind. The selective removal of fine-grained particles often leaves surfaces covered with coarse rocky particles known as **desert pavement**.

Sand, transported by the wind, is deposited to form **dunes**. Large areas of dunes are known as **sand seas**. Mobile dunes are called **barchans** or **crescentic dunes**. They have steep slip faces and lower windward slopes, with 'horns' pointing in the direction of the prevailing wind. **Linear dunes** are normally more than 100km long with slip faces on alternate sides.

Despite the low rainfall, temporary streams and rivers (**wadis**, **arroyos**) are powerful landforming agents. High intensity rainfall and the lack of plant cover mean that runoff is rapid, and streams and rivers carry very high sediment loads. Features such as **alluvial fans**, **bajadas**, **playas**, **pediments** and **canyons** are the result of fluvial processes.

Why are hot arid and semi-arid environments considered to be fragile?

Key ideas	Content
Climatic extremes lead to finely balanced ecosystems that can be easily damaged	The study of one hot arid/semi-arid environment (e.g. southwest USA) to illustrate: • the impact of climate on the nature of the ecosystem • how physical and human factors make the environment ecologically vulnerable
Both flora and fauna can suffer as a result of change, and regeneration is difficult in the harsh conditions	

Key questions

With reference to one specific hot arid/semi-arid environment, explain how climate influences the nature of the ecosystem.

Large parts of the US southwest, including the Mojave and Sonoran deserts, are arid or semi-arid. They are harsh environments for plants and animals because of their sparse rainfall, high rates of evaporation, very high summer temperatures and intense solar radiation. Few plants and animals have adapted successfully to these harsh conditions. As a consequence, primary production and biodiversity are low, and food webs are short.

Most endemic desert plants and animals in the US southwest have evolved special adaptations to the climate. **Succulents** like cacti (e.g. saguaro cactus, prickly pear) have thick waxy skins to seal in moisture, fleshy stems to store moisture, and spines to reduce transpiration. Joshua trees have needle-shaped leaves to reduce moisture loss, and creosote bushes (**phreatophytes**) send out long tap roots to absorb soil moisture. Tamarisk has long roots that reach down to the water table.

Some plants, such as the saltbush, tolerate high salt levels that would be toxic to other species. Annual plants avoid the effects of drought by completing their life cycles (germination, flowering, seeding) in just a few weeks after rain.

Animals have also adapted to the climate. Some have physiological adaptations. For example, jack rabbits have enormous ears to dissipate body heat, desert tortoises obtain all the moisture they need from their food, and others such as kangaroo rats can recycle water from their own urine.

Other animals cope with heat and water loss by adapting their behaviour. Animals that live in the desert all year round (e.g. reptiles and mammals such as bats and small rodents) are often nocturnal. Others, like the round-tailed ground squirrel, enter a state of hibernation (known as **aestivation**) during the hot season.

Knowledge check 15

Outline, with examples, the physiological adaptations made by plants to overcome problems of water shortage in desert environments.

With reference to one hot arid/semi-arid environment, explain how physical and human factors make the environment ecologically vulnerable.

Hot arid and semi-arid environments are finely balanced and easily degraded by human activities. Recovery from environmental damage can take decades. There are several reasons why hot arid and semi-arid environments are ecologically vulnerable:

- Lack of biodiversity means there are few negative feedback loops to buffer change.
- Shallow and dry soils are quickly eroded by wind and water if the vegetation cover is disturbed.
- Low rainfall, high temperatures and the low rates of plant growth make recovery from disturbance extremely slow.
- Plants and animals are highly specialised and unable to adapt to sudden environmental change.

In the deserts of the southwest USA, **cryptobiotic crusts** of algae, lichen, mosses and fungi form a fragile ground cover and are crucial to the health of entire ecosystems. Desert ecosystem depends on this crust because it:

- binds loose soil particles together
- absorbs and stores rainwater
- slows runoff
- inputs organic matter to the soil

Cryptobiotic crusts are easily damaged by hikers, mountain bikes, vehicles and domestic livestock. Once damaged, even small areas of crust can take up to 250 years to recover. Damage often triggers soil erosion, which in turn buries neighbouring crusts and prevents photosynthesis. In fragile desert ecosystems, once disturbed, large areas of crust (on which so many organisms depend) never recover.

Examiner tip

Describing the fragility of desert ecosystems in general terms is relatively straightforward. However, quality answers will be securely based in the context of actual places, ecosystems and processes.

What are the issues associated with the development of hot arid and semi-arid environments?

Key ideas	Content
Hot arid and semi-arid environments provide opportunities and challenges for development. Opportunities include: • resource exploitation, including agriculture, recreation and tourism Challenges include: • environmental constraints • costs/remoteness • conflicts with indigenous populations	The study of two contrasting hot arid/semi-arid environments (e.g. Draa Valley, Arches National Park) to illustrate: • the ways in which hot arid/semi-arid environments provide economic opportunities such as resource exploitation and recreation and tourism • the ways in which the development of hot arid/semi-arid environments present social, economic and environmental challenges, including: conflicts with indigenous populations; costs of development; environmental impacts

Key questions

In what ways do hot arid/semi-arid environments provide economic opportunities for development?

Hot arid environments such as the Draa Valley in Morocco provide opportunities for agriculture and tourism. Providing that water is available for irrigation, temperatures permit year-round cultivation. Irrigation water from the River Draa and from groundwater sources allows around 43,000ha to be cultivated and supports 225,000 people. Traditional polyculture — where cereal crops are grown alongside citrus and date palms — is sustainable, despite the fragility of the environment. In recent years, tourism has developed as an alternative source of employment. The region is easily accessible from Europe and offers a dry, sunny climate, access to the Sahara Desert, plus a distinctive Berber culture and vernacular architecture.

Examiner tip

When planning an answer to a question of this type start by listing possible resources and economic opportunities in arid and semi-arid environments (e.g. mining, energy supply, agriculture, tourism). Then relate each activity to examples you have studied.

The Arches National Park in Utah is a desert environment and has the highest concentration of natural arches in the world. The park receives nearly a million visitors a year. In addition to the stunning desert scenery, tourists come to experience the park's arid wilderness through hiking, camping and backpacking.

Hot arid environments also provide economic opportunities for the generation of solar energy. In the Mojave Desert in southern California, conditions are ideal (360 days of sunshine a year). The region already has nine solar power stations, and California's state government has plans to increase capacity to 3000MW in the next few years.

How do hot arid/semi-arid environments present social, economic and environmental challenges to development?

Fragile hot arid and semi-arid environments pose particular challenges to development. In truly arid environments, cultivation relies entirely on irrigation.

However, mismanagement through over-irrigation often causes problems. In Khushab, in northern Pakistan, **over-irrigation** has led to rising water tables and waterlogging. Evaporation then concentrates salts at or near the surface. The resulting **salinisation** severely reduces crop yields and can ultimately cause the abandonment of farmland (see Figure 7). In semi-arid regions such as Korqin in central China, deforestation by overgrazing and the clearing of woodland for cultivation, exposes dry soils to wind erosion and land **degradation**.

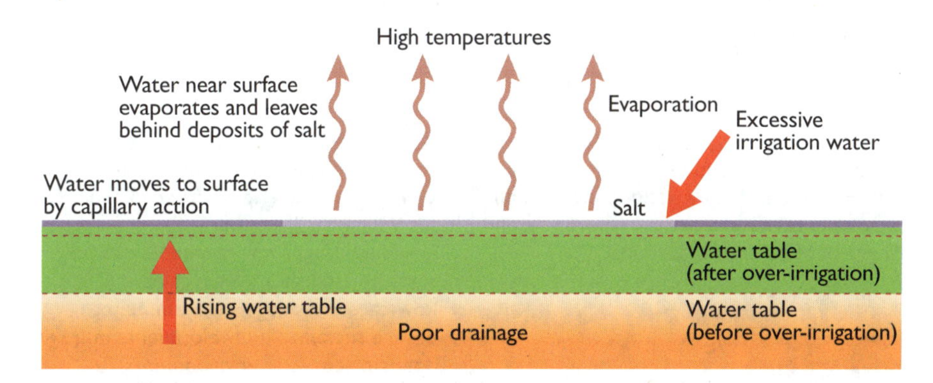

Figure 7 The salinisation process

Rainfall variability is high in hot arid and semi-arid environments and drought is common. In these circumstances, indigenous people often over-exploit pasture, fuelwood and water resources. In recent years in Nara, Mali, a combination of drought and population growth has resulted in severe land degradation and **desertification** (see Figure 8). This has had serious social and economic consequences, including a rise in poverty, food shortages, famine and malnutrition, and out-migration among indigenous groups.

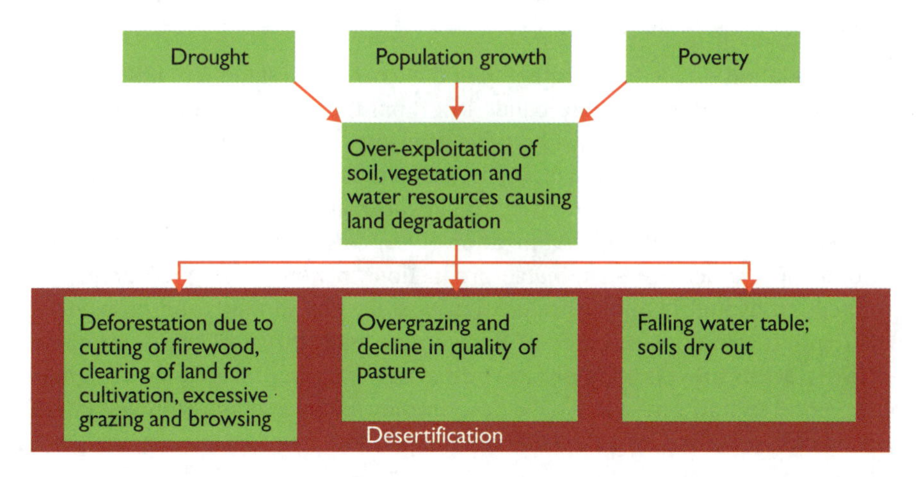

Figure 8 Causes of desertification in Nara, in the African state of Mali

In Morocco's Draa Valley there is a delicate balance between the supply and the demand for water. Increasing demand is leading to water shortages, while over-exploitation of groundwater has caused a decline in water quality. Both trends threaten to undermine the local farming economy. Meanwhile, salinisation and land

Knowledge check 16

What are the causes of land degradation and desertification in arid and semi-arid environments?

degradation remain urgent problems. In popular national parks like Arches in the USA, the large number of visitors puts increasing pressure on the environment, with permanent damage to cryptobiotic crusts and accelerated erosion on many trails.

How can hot arid and semi-arid environments be managed to ensure sustainability?

Key ideas	Content
Managing hot arid/semi-arid environments is often about balancing socioeconomic and environmental needs — this requires careful management to ensure sustainability	The study of two contrasting hot arid/semi-arid environments (e.g. northern China, Indus Valley) to illustrate: • how such fragile environments can be exploited for short-term gains • how careful management can help to ensure sustainable developments in fragile hot arid/semi-arid environments

Examiner tip

Definitions of key technical terms used in an essay title (e.g. sustainable development) should always be given in the opening paragraph of an answer.

Key question

With reference to two contrasting examples, explain how careful management can help to ensure sustainable development in hot arid/semi-arid environments.

Over-exploitation of resources and mismanagement by farmers has caused serious environmental damage in semi-arid northern China and the Jhelum Valley in northern Pakistan.

Land degradation — due to overgrazing and clearing woodland for cultivation — is widespread in the Korqin Sandy Lands in China. Recent increases in the demand for water have lowered the water table, causing further degradation. Low soil fertility, wind erosion, and salinisation are symptoms of land degradation. The Great Green Wall programme, sponsored by the governments of China and Belgium and by the Food and Agriculture Organization (FAO), is an attempt to rehabilitate the environment and achieve sustainable production. In Korqin, the programme aims to protect cultivated land against erosion, restore soil fertility, and provide local people with a sustainable source of timber. The main thrust of the programme is **reafforestation** and planting **shelterbelts**. New farming techniques designed to conserve soil fertility and water have been introduced, including controlled grazing, agro-forestry, recycling organic material, and planting tree species that can provide fodder crops.

Knowledge check 17

What are the environmental advantages of sustainable development?

Land degradation due to waterlogging and salinisation of soils is the main problem at Khushab in the Jhelum Valley. It has reduced crop yields, which has increased poverty and unemployment. In the worst affected areas, farmland has been abandoned. The World Bank sponsored a project to reclaim salinised land in the valley, which aimed to increase production and incomes, and achieve sustainability. Waterlogging and

salinisation were tackled by installing field underdrainage using PVC pipes, building new surface water drains, and lining irrigation canals to reduce water losses through seepage. The project was highly successful. Groundwater levels fell and within the first 2 years the waterlogged area was halved. The economic impact was immediate: farmers' incomes rose significantly and the average family income is now 34% above the official poverty level.

Summary

- Distinctive geomorphological processes operate in hot arid and semi-arid environments. These processes include salt and freeze–thaw weathering; wind (aeolian) abrasion, deflation and saltation; and water (fluvial) erosion, transport and deposition.
- The interaction of geomorphological processes with the geology and structure of arid and semi-arid environments produces characteristic landforms such as dunes, canyons, sculptured rocks, alluvial fans and playas (salt pans).
- Water has played an important role (today and in the past) in the evolution of physical landscapes in hot arid and semi-arid environments.
- Hot arid and semi-arid environments are regions of climatic extreme, with severe limits on biodiversity, biomass and plant productivity. In these conditions ecosystems are fragile and sensitive to human activity. Following human disturbance ecosystems are slow to recover and may, in the long-term, suffer permanent degradation.
- Hot arid and semi-arid environments provide opportunities for economic development through mineral and energy resource exploitation, agriculture and tourism.
- Environmental harshness, remoteness, costs and conflict with indigenous people provide obstacles to development in many hot arid and semi-arid regions. In the past development has often resulted in land degradation (e.g. soil erosion, salinisation, deforestation) to the point where land is abandoned.
- Development in hot arid and semi-arid environments requires careful management to ensure long-term sustainability. Planning is needed to prevent over-exploitation of physical and biological resources, which leads to soil exhaustion, soil erosion, salinisation, overgrazing, deforestation and so on.

Questions & Answers

Assessment

F761 Managing Physical Environments is one of two units that make up the AS specification. It is worth 100 uniform marks, and accounts for 50% of the specification weighting.

The other unit is F762 Managing Change in Human Environments (see Table 1).

Table 1 AS Geography: scheme of assessment

Unit number and unit name (exam length)	Raw marks	Uniform marks (AS weighting)
F761 Managing Physical Environments (1½ hours)	75	100 (50%)
F762 Managing Change in Human Environments (1½ hours)	75	100 (50%)

Unit F761 covers four major topics:

- River environments
- Coastal environments
- Cold environments
- Hot arid and semi-arid environments

The exam paper is in two parts — Section A and Section B.

In Section A you must answer **two** data-response questions, one from **either** River environments **or** Coastal environments, and one from **either** Cold environments **or** Hot arid and semi-arid environments.

Section B requires you to answer **one** extended-writing question. One question is set on each of the four topics. The question you select in Section B must be on a different topic from the two you chose in Section A. The requirement of the question paper means, therefore, that you must study **at least three** of the four topics specified for Unit F761. To maximise your choice of question you would need to study all four topics.

Data-response questions

The data-response questions in Section A are divided into four sub-questions (a(i), a(ii), b, c) worth 4, 6, 6 and 9 marks respectively. Two of these sub-questions (b, c) usually require knowledge of one or more geographical examples, and two (a(i), a(ii)) are usually linked to stimulus materials such as OS maps, charts, photographs and diagrams.

Data response questions are worth 50 out of the 75 marks available for Unit F761. Thus, in a 1.5 hour exam, you should devote approximately 30 minutes to each structured question.

Extended-writing questions

Section B requires you to answer an extended-writing or essay-style question. You should allow approximately 30 minutes for this. These questions demand description, analysis, application — and, most importantly, detailed reference to examples and case studies.

Mark scheme criteria

Examination answers are assessed against three criteria or assessment objectives (AOs). For AS Geography these are as follows:

- **AO1 Demonstrate knowledge and understanding** of the specification content, concepts and processes.
- **AO2 Analyse, interpret and evaluate** geographical information, issues and viewpoints, and apply them in unfamiliar contexts.
- **AO3 Investigate, conclude and communicate**, by selecting and using a range of methods, skills and techniques to investigate questions and issues, reach conclusions and communicate findings.

You are advised to study these criteria carefully because they tell you how your examination answers will be judged. The section on examination skills below explains how assessment objectives are used in mark schemes. Table 2 shows the weighting given to each AO.

Table 2 Assessment objective weightings in AS Geography

Unit number and unit name	% of AS			
	AO1	AO2	AO3	Total%
AS Unit F761 Managing Physical Environments	25	10	15	50
AS Unit F762 Managing Change in Human Environments	25	10	15	50

Examination skills

Answering data-response questions

Data-response questions have a gradient of difficulty. The initial questions are less demanding and carry fewer marks than later questions. Typically the first one or two questions use command words such as 'describe' or 'outline', while later questions require explanations illustrated by geographical examples.

Stimulus materials are used both directly and indirectly. For direct use, OS maps, satellite images and photographs are provided to assess geographical skills, such as map reading and interpretation, while charts and sketch maps test your ability to summarise, recognise, describe and analyse spatial patterns and trends. For indirect use, stimulus materials provide a catalyst for assessing wider knowledge and understanding of a topic.

Mark schemes for data-response questions are levels-based. For example, there are two levels of attainment for 4- and 6-mark questions, and three levels for 9-mark questions. Marks tend to be weighted towards the top end. For example, in a 9-mark question, 8 or 9 marks will be reserved for a level 3 answer. Each level is defined by descriptors (see Table 3). Having read an answer, examiners first assign it to a level, and then decide the precise mark within that level.

Table 3 Basic descriptors used to assess structured questions

4-mark questions		
Level	**Mark**	**General descriptor**
2	3–4	Identifies relevant features/processes etc.; clear descriptions; cause–effect explained; good technical language
1	0–2	Identifies a relevant feature/process etc.; basic or no description; links often unexplained; gaps in technical language
6-mark questions		
Level	**Mark**	**General descriptor**
2	5–6	Clear understanding; detailed explanation and links fully explained; good technical language
1	0–4	Limited understanding; links not fully explained; gaps in technical language
9-mark questions		
Level	**Mark**	**General descriptor**
3	8–9	A good range of clear, detailed and valid reasons/factors/causes etc. explained; well-chosen examples; well-structured; accurate grammar and spelling; accurate terminology; uses clearly identified example(s)
2	5–7	A number of valid reasons/factors/causes etc. though explanation may not be clear; some examples; sound structure; some errors of grammar and spelling; some use of appropriate geographical terms; gives clearly identified example(s)
1	0–4	Some valid reasons/factors/causes; descriptive with little or no reference to cause–effect relationships; limited or no examples; little structure; some errors of grammar and spelling; little use of geographical terms

When answering structured questions, you should follow these guidelines:

- Read all parts of the question before attempting to answer. This will help you to avoid repetition in later answers and allow you to get an overview of how the topic is developed.
- Study the stimulus material carefully.
- Make sure you understand precisely what each question is asking you to do.
- For the 9-mark question, which may require up to 25 lines of writing, plan your answer. You can do this by making a list of the key points and the specific examples you want to include in your answer.
- Divide your time realistically and adjust the length of your answers to the mark weighting. Six lines of writing are probably sufficient for a 4-mark question, whereas for a 9-marker you will need to write at least three times as much.

Answering extended-writing questions

Answers to extended-writing questions are assessed against three criteria (or assessment objectives). Each assessment objective is divided into three attainment levels, with a maximum of 13 marks for knowledge and understanding, 5 for analysis and application, and 7 for skills and communication (see Table 4). The relatively large weighting given to skills and communication underlines the importance of accurate spelling and grammar, as well as the ability to structure an answer and provide a clear conclusion.

Table 4 Basic descriptors used to assess extended-writing questions

Knowledge and understanding		
Level	**Mark**	**General descriptor**
3	11–13	Detailed knowledge and understanding; cause–effect links are clearly explained; there is effective use of detailed exemplification
2	7–10	Some knowledge and understanding; cause–effect links are stated but not clearly explained; cause and effect is understood and there is use of exemplification
1	1–6	Limited knowledge and understanding; no cause–effect links are stated and there is limited exemplification; if no located example then top of Level 1 maximum
Analysis and application		
Level	**Mark**	**General descriptor**
3	5	Clear analysis and application of knowledge and understanding to the demands of the question
2	3–4	Some analysis and application of knowledge and understanding to the demands of the question
1	0–2	Limited analysis and application of knowledge and understanding to the demands of the question
Skills and communication		
Level	**Mark**	**General descriptor**
3	6–7	Answer is well structured; accurate use of grammar and spelling; good use of appropriate geographical terminology; clear conclusions
2	4–5	Answer has sound structure but may have some errors in grammar and spelling; some use of appropriate geographical terminology; conclusions are attempted
1	0–3	Answer has little structure and has some errors in grammar and spelling; little use of appropriate geographical terminology; no conclusions are attempted

All extended-writing questions are based on the content of the last two 'questions for investigation' for each topic in the specification. As a result, they focus mainly on geographical problems, opportunities and management responses.

Extended-writing questions have a number of common features:

- They require description *and* explanation.
- They always require exemplification using at least two (often contrasting) located case studies.
- They include a brief introduction and a conclusion.

Command words and phrases

Command words and phrases in examination questions are crucial because they tell you exactly what you have to do. You must respond precisely to their instructions. For example, the instruction 'describe' is very different from 'explain'. Ignoring command words and phrases is a common error, and is a major cause of under-achievement. Table 5 lists some typical command words and phrases used in questions in the OCR AS Geography examination and explains what they require you to do.

Table 5 Key command words and phrases

Command word/phrase	Requirements
Describe	Provide a word picture of a feature, pattern or process. Descriptions in short-answer questions are likely to be worth 4 or 6 marks and will require some detail.
Outline	The same as 'describe' but requiring less detail. The idea is to identify the basic characteristics of a feature, pattern or process.
Compare	Describe the similarities and differences of at least two features, patterns or processes.
Examine	Describe and comment on a pattern, process or idea. 'Examine' often refers to ideas or arguments that demand close scrutiny from different viewpoints.
Why?/Explain/Account for/ Give reasons	Give the causes of a feature, process or pattern. This usually requires an understanding of processes. Explanation is a higher-level skill than description and this is reflected in its greater mark weighting in examination questions.
Identify	State or name features, processes, landforms etc.
Suggest	Give possible or likely reasons. The reasons need to be plausible rather than definitive.

Case studies

An important feature of the OCR AS Geography specification is its emphasis on exemplification through detailed case studies. All the extended-writing questions in Section B, and the 9-mark structured question in Section A, require examples that refer to specific geographical areas. These questions deal mainly with environmental, social and economic problems — and management responses to them.

In addition, the extended answer questions often require examples that demonstrate the importance of sustainable approaches to management. Generalised answers to these questions will not achieve the highest level. Your revision of content for each topic must, therefore, always include at least one, and sometimes two, case studies.

Student Questions & Answers section

This section contains examples of student answers to eight examination questions, covering the four topic areas outlined in the Content Guidance: River environments, Coastal environments, Cold environments, and Hot arid and semi-arid environments.

Section A contains four data-response questions (one for each option), and Section B has four extended-writing or essay questions (again, there is one question for each option).

Each data-response question consists of four sub-parts and is constructed around stimulus materials such as OS maps, sketch maps, diagrams, charts and photographs.

In the examination, you will have to answer two data-response questions and one extended-writing question. You should give yourself around 30 minutes for each question. The lengths of your answers to the structured questions should be proportional to the mark weighting, which varies from 4 to 9 marks. A rough guide is to think in terms of 1 mark for every 1.5 to 2 lines of writing.

Examiner's comments

Examiner comments on the questions are preceded by the icon e. They offer tips on what you need to do in order to gain full marks. All candidate responses are followed by examiner's comments, indicated by the icon e, which show how marks have been awarded and highlight areas of credit and weakness. For weaker answers the comments suggest areas for improvement, by highlighting specific problems and common errors such as lack of development, excessive generalisation and irrelevance.

Section A: structured questions

Question 1 River environments

Michael Raw

Figure 1 Floodplain of the Marshaw Wyre, north Lancashire

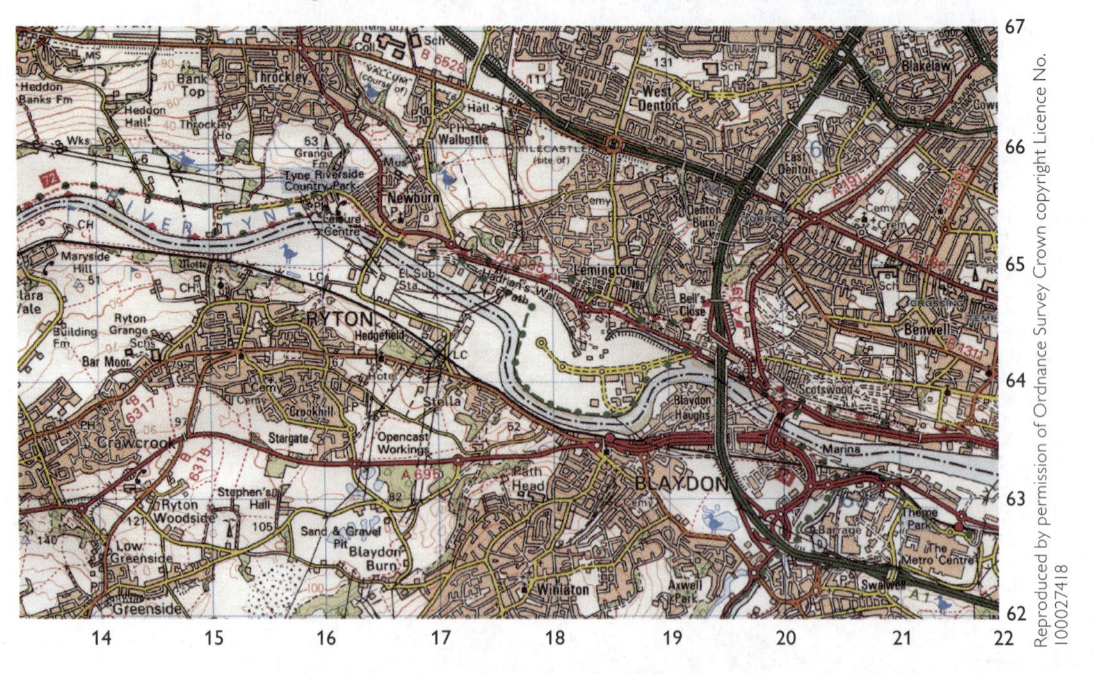

Reproduced by permission of Ordnance Survey Crown copyright Licence No. 100027418

Figure 2 OS map of part of Newcastle-upon-Tyne

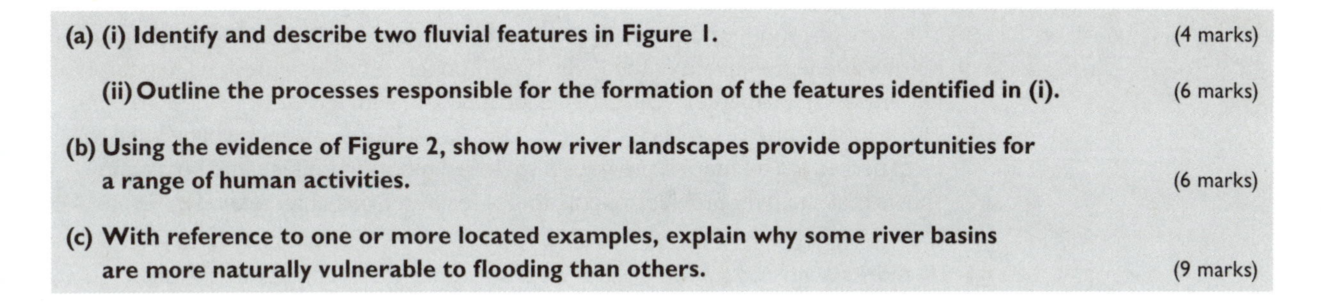

(a) (i) Identify and describe two fluvial features in Figure 1. (4 marks)

(ii) Outline the processes responsible for the formation of the features identified in (i). (6 marks)

(b) Using the evidence of Figure 2, show how river landscapes provide opportunities for a range of human activities. (6 marks)

(c) With reference to one or more located examples, explain why some river basins are more naturally vulnerable to flooding than others. (9 marks)

e Note how the questions have a gradient of difficulty from simple statements and descriptions requiring knowledge in (a) to explanation in (c). The marks allocated to each part question reflect the levels of difficulty of these demands.

D-grade answer

(a) (i) Two fluvial features in Figure 1 are: a point bar and a floodplain **a**. The point bar is where sediment is dropped due to the river not being strong enough to carry it **b**. The floodplain is where over many years the river has changed course and created this area **b**. The river cliff is found where the river meets the valley side and erosion widens the floodplain **b**.

e **2/4 marks awarded** **a** The answer correctly identifies two features from the photograph and is awarded 2 marks. **b** However, instead of describing the features, the candidate attempts to explain them. This explanation is irrelevant and scores no marks. The answer achieves Level 1 for correctly identifying two features.

(ii) The point bar is formed as the river meanders. The current is fastest on the outside of the meander bend and slowest on the inside **a**. This means that the transport of sediment will be weaker on the inner bend so sediment gets left **b** on the edge **a** of the channel creating a point bar. The floodplain is formed by the river changing its course and expanding the sides of **b** its valley by lateral erosion **a**. The floodplain is made up of sediment dropped by **b** the river **a**. Also when the river bursts its banks it leaves a layer of sediment on the valley floor.

e **3/6 marks awarded** **a** The answer shows a lack of clear understanding of the processes involved in the formation of both features. For example, the importance of helical flow in the point bar formation and the significance of suspension load and bedload in the alluvial fill on the floodplain are not mentioned. The rather sketchy understanding of processes merits just 3 marks out of 6. **b** There are also limitations in the use of technical language.

(b) The area surrounding the river provides a bird sanctuary, a leisure centre and a marina **a**. The floodplain of the River Tyne probably provides a wetland habitat for birds **a**. The sanctuary will provide local people with a chance to view wildlife. The leisure centre occupies a flat site on the floodplain and again will be used by local people **b**. The marina provides a sheltered mooring for pleasure boats and shows that the river provides opportunities for water-based recreation **c**.

2/6 marks awarded **a** A weakness in the answer is the lack of any specific reference to the OS map. **b** Although the answer correctly identifies a number of relevant activities, the explanations are superficial. **c** The answer focuses exclusively on recreation and leisure, ignoring a whole range of other activities such as industry, power stations, water supply and water treatment and so on. Overall this is a Level 1 answer.

(c) A number of natural factors (not man-made) influence flooding, such as geology, slopes, vegetation and rainfall. In July 2006 the River Calder in West Yorkshire caused flooding in Hebden Bridge **a**. This was mainly due to very high rainfall in a short period of time on the Pennines **b**. Other reasons for flooding were the steep slopes of the river basin, the impermeable rocks (e.g. gritstone and shale) and the small amount of woodland cover **b**.

Urbanisation often makes rivers more prone to flooding **b**. This is because towns and cities have rapid runoff **c** due to impermeable tarmac and concrete surfaces, and sewerage systems. The floods in Sheffield in summer 2007 when the River Don overtopped its banks show this **a**.

In southern France, the Laval and Brusquet river basins are situated close together **a**. However, the Laval is much more likely to flood than the Brusquet. This is explained by the fact that 87% of the Brusquet river basin is covered in forest, compared to only 22% in the Laval river basin **a**. In the Laval basin, rainfall runs quickly into streams and rivers but in the Brusquet basin large amounts of rainfall are intercepted by trees. This causes much of the rainwater to be evaporated **b**, reducing the risk of flooding **c**.

7/9 marks awarded **a** The examples used in the answer are well chosen, with the Calder and Laval-Brusquet examples contrasting well although a little more detail could have been provided. **b** The answer makes clear the distinction between natural and man-made factors involved in flooding. **c** The answer addresses the question and avoids irrelevant descriptions of flood control features such as levées and dams. The use of terminology, grammar and spelling is sound. Overall this a good Level 2 answer, which with a little more development could have achieved Level 3.

The total score for this four-part answer is 14/25, a D grade.

A-grade answer

(a) (i) A floodplain and a point bar are two fluvial features in the photograph **a**. The floodplain is the flat valley floor on either side of the river channel. It is probably 50–100m wide. Steep slopes mark the boundary of the floodplain **b**. The point bar is the accumulation of pebbles and cobbles on the inner bank of the meander's channel **b**.

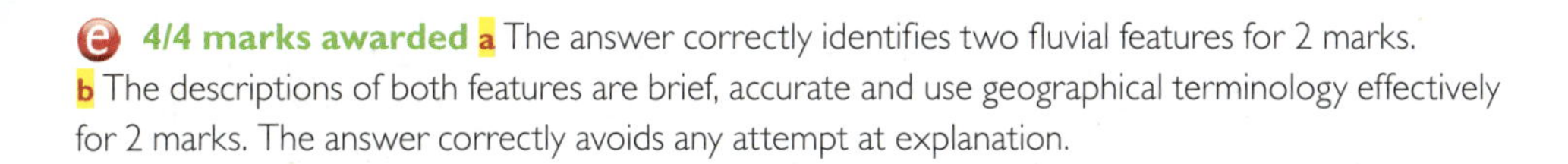
4/4 marks awarded **a** The answer correctly identifies two fluvial features for 2 marks. **b** The descriptions of both features are brief, accurate and use geographical terminology effectively for 2 marks. The answer correctly avoids any attempt at explanation.

(ii) Floodplains are formed by erosional and depositional processes. As meanders move across the valley floor (due to lateral erosion **a**) they undercut the valley sides and widen out the valley. Meanwhile, sediment deposited in the channel (e.g. as point bars) is abandoned and silt and clay are deposited across the valley floor during floods. All of this sediment (known as alluvium **a**) fills in the valley floor and helps to build up the floodplain **b**. Point bars are depositional features that form on the inner bank of meanders. During high flow a current moves close to the river bed towards the inner bank (helical flow **a**), depositing first coarse, then finer sediment. Gradually this results in a point bar **b**.

6/6 marks awarded **a** Accurate terminology is used appropriately throughout the answer. **b** The answer demonstrates a detailed and accurate understanding of fluvial processes and links process and form successfully.

(b) The main human activities on the map which are linked to the River Tyne are: a marina (2063), a country park (1665), a long distance footpath (Hadrian's Wall footpath), an electricity sub-station (1764) and a number of works/factories (1864) **a**. The river provides opportunities for recreation and leisure activities. A marina suggests that sailing is important. The country park shows how rivers offer attractive environments for wildlife and for recreation. The long distance footpath probably takes advantage of the flat land and open space alongside the river **b**. The river is bordered by a wide floodplain. This provides large flat sites for factories and commercial buildings.

5/6 marks awarded **a** Specific references to the map, using grid references and the names of features, satisfy the instruction to use 'the evidence of Figure 2'. **b** Several valid land uses and opportunities are identified, though there could be more development of the reference to flat sites and land uses unaffected by flooding. In general this is a relevant answer to the question, which achieves Level 2.

(c) Some river basins are naturally more prone to flooding than others **a**. Among the natural factors which affect flooding are: vegetation cover, slopes, rock type and climate. The first three factors affect lag times and peak flows **b**. Generally the shorter the lag time the higher the peak flow and the more likely that flooding will take place.

In the southern Lake District the River Kent has short lag times and high peak flows, and floods often threaten parts of the market town of Kendal. Several factors contribute to the River Kent being a 'flashy' river. Its headwaters are the mountains in southern Lakeland, over 800m high **c**. The slopes are steep and the rocks are impermeable grits, tuffs and lavas. This means rapid runoff and high

peak flows. In addition, the mountains have only sparse vegetation cover with few trees, reducing interception and transpiration **b**. Finally, the mountains often receive torrential rain and get well over 2500 mm of rain a year. All this combines to create a high flood risk and especially in the winter months if there is melting of lying snow.

The River Test in Hampshire is, by contrast, far less prone to flooding. Slopes in its drainage basin are fairly gentle and the fall in level from the headwaters on Salisbury Plain to the river's mouth in the Solent is less than 300 m. Although there is no more tree cover than in the Kent's drainage basin, chalk is the main rock type in the catchment **c**. Chalk is a permeable rock. It absorbs and stores water. It is a huge underground reservoir that releases water slowly, making the flow of the Test very stable. The regular flow of the river makes flooding far less likely than on the River Kent. Finally, the Test drains one of the driest areas in England, with rainfall less than 700 mm and evaporation and transpiration **b** high in the summer.

8/9 marks awarded **a** Some definition of what is meant by 'naturally vulnerable' to flooding would normally help to clarify the question and avoid possible irrelevance. **b** The answer is well structured and there is accurate spelling, grammar and use of terminology. **c** Knowledge and understanding of two specific catchments is quite detailed and has been applied relevantly. The answer does not stray into areas of irrelevance such as flood management and human factors that exacerbate floods (e.g. urban land use, deforestation). This is a solid Level 3 answer.

The total score for this four-part answer is 23/25, an A grade.

Question 2 **Coastal environments**

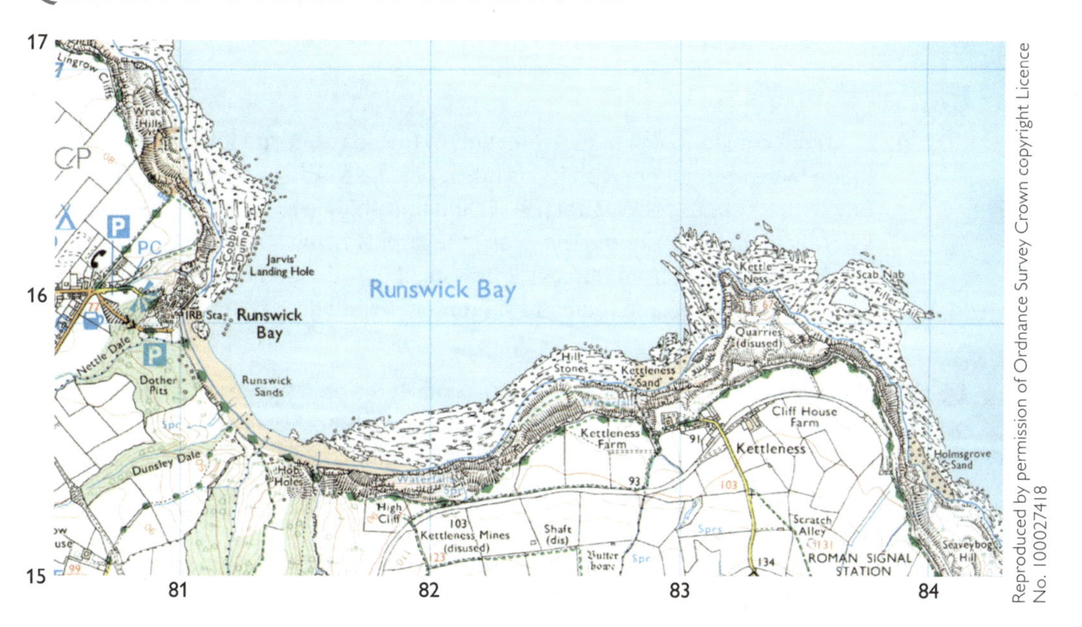

Figure 3 OS map extract of Runswick Bay, North Yorkshire

Figure 4 Hard-engineered coastal protection at Overstrand

(a) (i) Describe the main physical features of the coastline in Figure 3. (4 marks)

(ii) Outline the formation of the physical features described in (i). (6 marks)

(b) Using Figure 4, explain how coastal protection can take a variety of forms. (6 marks)

(c) With reference to one or more located examples, explain why coastal management increasingly favours soft engineering and managed retreat. (9 marks)

Read both parts of question (a) before starting to answer. This is because questions (i) and (ii) are linked, and accuracy is needed to differentiate between description (required in (i)) and explanation (required in (ii)).

C-grade answer

(a) (i) From the constantly changing direction of the coastline in Figure 3, it is clear that a huge amount of erosion has taken place. Erosion has created Runswick Bay with its small pocket beach **a** of sand and shingle **b**. There are steep cliffs **a** nearly all the way around the coast, the largest being Lingrow Cliffs in the northwest **b**. Apart from Runswick Bay, most of the coastline is occupied by a shore platform **a**, **b**, and there is a prominent headland **a** at Kettleness.

2/4 marks awarded **a** This answer correctly identifies the main features along this stretch of coastline, though the **b** descriptions of the features are limited. More detailed description and specific reference to the map (e.g. height of cliffs, direction of the coastline, width of the shore platform) are needed for 3 or 4 marks.

(ii) The formation of Runswick Bay **a** is probably due to the erosion of powerful waves from the north **b**. The rocks in Runswick Bay may be weaker than those to the east and west, and therefore are more easily eroded. Between the bays **a** the rocks are more resistant and do not erode so easily as headlands **c**. This leads to the formation of headlands like Kettleness. Cliffs form as a result of wave action (abrasion, hydraulic action) forming a notch at the cliff base. As the cliffs are undermined they collapse and slowly retreat inland **c**. Shore platforms **b** are exposed at low tide along much of the coastline **a** in Figure 3. They represent the base of old cliffs eroded into caves, arches, stacks, stumps and eventually into gently sloping, rocky platforms.

The small pocket beach at Runswick **a** has been deposited by wave action. It is roughly crescent-shaped and therefore seems to have formed parallel to the waves. It is therefore swash-aligned, which means that the waves in the bay will be fully refracted **c**.

5/6 marks awarded **a** The answer could include more references to the specific characteristics of the coastline shown by the map (e.g. grid references, place names). **b** The answer lacks some development in places (e.g. direction of dominant waves), but **c** it shows a clear understanding of coastal processes. Overall, this answer deals accurately (and in some detail) with the major features of the coastline.

(b) Figure 4 shows three different forms of hard coastal defences: groynes, a sea wall and gabions **a**. Groynes protect the coast from erosion by cutting out longshore drift. This builds up beaches that reduce erosion as wave energy is absorbed by the sand before reaching the cliffs **b**. The seawall is a concrete wall at the base of the cliffs. It reflects wave energy and helps to prevent erosion **b**. Seawalls are, however, rather unsightly. Gabions are metal cages filled with rocks that break up waves and absorb wave energy **b**. Eventually the cages can get rusty and become sharp and dangerous.

3/6 marks awarded **a** This answer is largely a description of three types of coastal defence. **b** The explanation of how groynes function is thin but adequate, but the answer provides relatively little information on seawalls and gabions. For example, seawalls provide complete protection to a coastline but create their own problems such as basal scour (due to wave reflection), high costs of maintenance and construction. Seawalls are only justified where an important settlement or infrastructure is at risk from erosion. Gabions in this example are stabilising the coastal slope above the seawall and the high tide mark. The limited explanation provided in this answer results in a moderate Level 2 score.

(c) On the Norfolk coast at Happisburgh, a policy of 'no active' intervention and managed retreat is in operation **a**. Although controversial, this policy has several advantages. At Happisburgh, the cost of hard sea defences such as revetments would be a lot more than the properties at risk from erosion are worth **b**. Also with climate change and rising sea levels, hard engineering structures may be unsustainable **c** in the future **b**. A further problem is that when structures like seawalls and groynes are built, they reduce inputs of sediment **c** from erosion and stop the movement of sand along the coast. The effect is often to increase erosion further down the coast **b**. Eventually at Happisburgh erosion will cause a bay to form and erosion will stabilise naturally **c**.

Some people believe that we should not interfere with natural processes because it often results in further problems, and that nature should be left to take its course. But this is very controversial because it inevitably leads to homes and businesses being threatened. At Happisburgh an action group has been formed to persuade the government to fund coastal defences, and protect the village and people's homes.

7/9 marks awarded **a** Happisburgh provides a relevant example and context. **b** The answer correctly spells out three or four reasons in variable detail. The reasoning is sound. **c** The answer is structured and uses geographical terminology accurately. Ultimately the answer achieves the top of Level 2. For higher marks a fuller explanation of the effects of climate change and sustainability are needed.

The total score for this four-part answer is 17/25, a solid C grade.

A-grade answer

(a) (i) This is a rugged upland stretch of coastline **a**, with the land rising to nearly 90m **a**. It is dominated by erosional features **b**. Cliffs run along the entire length of the coast (apart from the western side of Runswick Bay) **a**. There is a headland at Kettleness and a classic bay at Runswick. Extensive areas of shore platform are found at the base of the cliffs, e.g. at Cobble Dump (8116) and Hill Stones (8215) **a**. The only significant depositional feature **b** is the crescent-shaped bay-head beach at Runswick **a**.

4/4 marks awarded **a** The answer accurately describes the main features of the coastline, and includes several clear references to map evidence. **b** The division of features into erosional and depositional types gives the answer a clear structure. This is a relevant Level 2 answer, which focuses wholly on description.

(ii) The cliffs are formed mainly by marine erosion **a**. Because this is an area of strong relief, the coast, eroded by wave action, forms high cliffs. The erosional processes at work are abrasion and hydraulic action **a**, which cut a notch **a** at the base of the cliffs. Eventually the cliffs are undermined, collapse and retreat inland. As they retreat they leave behind a gently sloping, rocky shore platform **b**. Apart from wave erosion, weathering **a** occurs on the shore platforms at low tide.

Headlands and bays often form where alternating bands of hard and soft rock meet the coast at right angles **b**. Erosion is faster on the softer rocks and they retreat rapidly to form a bay like Runswick. The more resistant rocks form headlands such as Kettleness **b**.

Runswick Bay is semi-circular in shape. This is due to the refraction **a** of waves entering the shallow waters of the bay **b**. The waves fan out in all directions and eventually break parallel to the coast. This process moves sand into the bay and is responsible for the crescent-shaped beach at the head of Runswick Bay.

6/6 marks awarded **a** The answer uses terminology accurately throughout. **b** There is detailed and convincing explanation of all the main features. Overall this is an excellent answer, which is well constructed and shows sound understanding of the relationships between physical processes and landforms.

(b) The coastal protection measures in Figure 4 are sea walls, gabions and groynes. Sea walls offer complete protection against erosion and flooding. They are vulnerable to scouring and undermining by wave action (which can lead to toppling) and the toe is often reinforced with metal piling and armour blocks. The upper part of sea walls is often concave in cross section. This reflects waves seawards and helps to prevent overtopping and flooding **a**,**b**.

In Figure 4, gabions seem to be used to strengthen the coastal slope above the sea wall and promenade. They consist of wire cages filled with rock particles and are a cheaper alternative to sea walls. Where they are used on the shoreline to prevent wave erosion they absorb wave energy in the spaces between the rock particles **a**,**b**.

Groynes are wooden fences that extend seawards from the high tide mark at right angles to the coast. They intercept sand and shingle being carried along the coast by longshore drift. As the sand accumulates behind the groynes it builds up a wide beach. This is ideal because beaches are very good at absorbing wave energy. Waves spend their energy on beaches by shifting sand and shingle rather than on erosion **a**,**b**.

e **6/6 marks awarded** **a** This answer provides accurate description of the three coastal protection structures and explains how they function. **b** Knowledge and understanding are applied relevantly throughout the answer. This is a detailed and accurate answer, which merits full marks.

(c) Coastal management increasingly favours soft engineering and managed retreat for a number of reasons. The first **a** is that the alternative — hard engineering — is expensive to build and maintain. Sea walls cost the most — £4000 to £7000 per metre. At Happisburgh in Norfolk **b** the sea defences (revetments) were badly damaged by a storm in 1990. The cost of rebuilding was not justified as it was greater than the value of the properties the sea defences protected. The result of this 'no active intervention' policy was rapid cliff erosion. Nature was left to take its course.

Second **a**, on a similar theme, hard sea defences will be more costly to maintain in future. This is explained by rising sea levels and increased storminess due to global warming and climate change. Hard defences like sea walls, revetments and groynes along some stretches of coastline are unsustainable.

Third **a**, hard structures interfere with natural coastal processes. Solving one problem often causes others. For example, the rock groynes at Mappleton on the Holderness **b** coast interrupt the flow of sediment along the coast. Beaches downdrift of the groynes have thinned since 1991 when the groynes were completed. The effect has been rapid erosion downdrift and the accelerated loss of farmland and even a farm at North Cowden.

Finally **a**, soft engineering, such as managed retreat, is favoured because it works with nature. Hard defences are replaced by mudflats and salt marshes (e.g. Blackwater estuary in Essex **b**). They renew themselves, provide protection from erosion and flooding, and provide valuable new habitats for wildlife.

e **8/9 marks awarded** **a** This detailed answer provides four clear explanations, which are clearly signposted. **b** Arguments are supported by actual examples, which are succinct and give the answer vital place specificity. Although there is scope for further explanation (e.g. the effects of seawalls on reducing sediment inputs to the coastal system, and the loss of salt marshes and mudflats due to 'coastal squeeze'), overall, this is a quality answer, showing good knowledge and understanding and their effective application.

e **The total score for this four-part answer is 24/25, a good A grade.**

Question 3 **Cold environments**

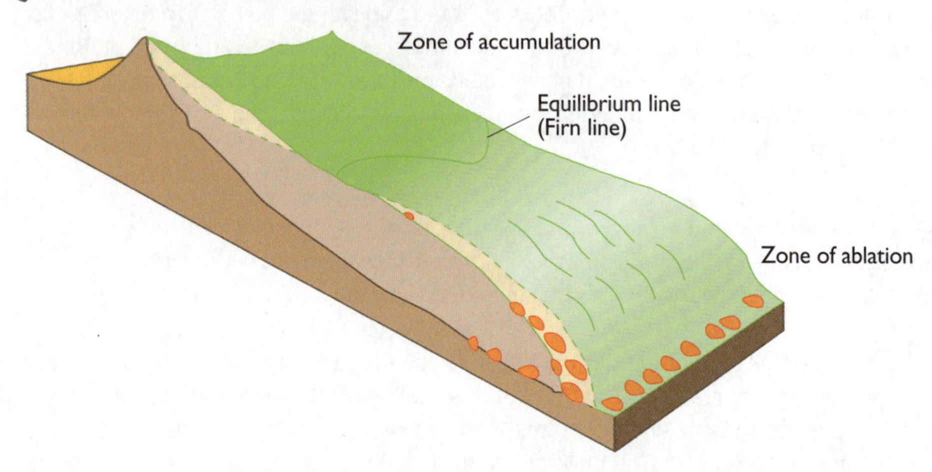

Figure 5 Valley glacier: zones of accumulation and ablation

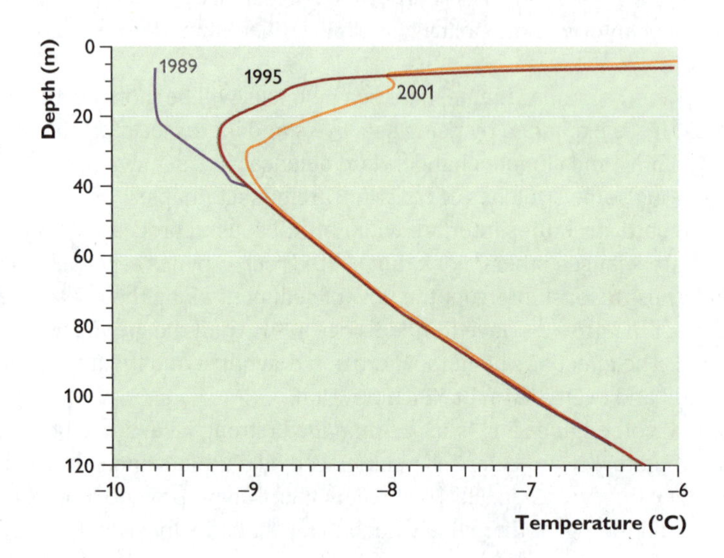

Figure 6 Permafrost temperature chart for Alaska

(a) (i) Using Figure 5, describe the main features of the mass balance of a glacier. (4 marks)
(ii) Show how changes in mass balance can affect the behaviour of glaciers. (6 marks)

(b) Explain the formation of the variety of morainic features that occur in glaciated lowlands. (6 marks)

(c) With reference to Figure 6, and one or more located examples, comment on the problems associated with development in cold environments. (9 marks)

e For (a)(i) it is important to study Figure 5 carefully and use the terms labelled on the diagram. Explanations of three or four different morainic features are required in question (b). This number allows both appropriate breadth of coverage and a reasonable amount of explanatory detail. In (c) the instruction 'comment on' invites to you briefly to describe, explain and offer evaluative insights into problems in cold environments.

C-grade answer

(a) (i) Mass balance refers to the annual input of snow to a glacier, and output by melting and evaporation. The main features of mass balance in Figure 5 are the zone of accumulation, where snowfall is greater than ablation **a**, and the zone of ablation, where ablation exceeds snowfall **a**. The equilibrium line is where snowfall and ablation are equal **a**.

3/4 marks awarded **a** This answer provides some interpretation of the features in Figure 5. It refers to the zones of accumulation and ablation, and to the equilibrium line. This is a Level 2 answer, but requires a little more descriptive detail to secure full marks.

(ii) Changes in mass balance can affect the direction a glacier moves in. If overall, the amount of snowfall is greater than the amount of ablation, a positive mass balance **a** is created and the glacier will advance **b** and continue to erode the landscape **c**. But on the other hand if the amount of snow and ice lost to ablation exceeds the annual accumulation **c** of snow and ice, the glacier will retreat and shrink **b**. In contrast to a positive mass balance, a negative balance **a** is likely to result in the deposition of moraine **a**,**c**.

4/6 marks awarded **a** There is accurate use of terminology. **b** The answer provides relevant explanation of the causes of glacial advance and retreat. **c** Some connections to glacial processes are made. The answer achieves a top Level 1 mark. Additional references to the concept of glaciers as systems, with inputs and outputs, and the specific circumstances that promote positive and negative mass balances would place the answer in Level 2. In summary, this is a brief but accurate answer.

(b) Rock debris from glacial erosion, weathering and mass movement is carried by the glacier. It is usually transported along the base or edges of the glacier. When deposited it is known as lateral and ground moraine respectively. Lateral moraines accumulate along the valley sides **a**. If a glacier remains stationary for a long period of time a moraine is deposited across the valley floor **b**. This is known as a terminal moraine **a**. During deglaciation, as a glacier retreats upvalley **b** it deposits small terminal moraines known as recessional moraines **a**. When two glaciers meet **b** the two lateral moraines join to form a medial moraine **a** running down the middle of the enlarged glacier.

3/6 marks awarded **a** The answer accurately describes several types of moraine, **b** but conveys only a superficial idea of the formation of moraines, and a limited appreciation of the processes involved. This is a Level 1 answer. It lacks a clear definition of moraines or any distinction between the moraines left by valley glaciers (e.g. lateral moraines, push moraines) and those left by ice sheets (e.g. till plains, drumlins); and it largely ignores explanation.

(c) In the late 1960s, oil and gas reserves were discovered in the Alaskan tundra **a**. Its development created many problems for the fragile tundra ecosystem **a** and the oil and gas companies had to find ways to extract the oil and gas without damaging the permafrost **b**. Despite precautions there have been problems with permafrost melting turning the surface into a morass. This can be seen in Figure 6 **c**, which shows the top layers of the permafrost getting warmer.

To prevent the permafrost from thawing from the heat generated by buildings and machinery, the oil and gas companies try to protect the permafrost by keeping a layer of insulating vegetation in place **d**. Other ways of protection include elevating buildings on poles and using thick layers of insulating gravel in foundations **d**.

The use of gravel as an insulator has led to other problems. Gravel extraction from river beds **a** has harmed fish populations and damaged river environments **b**. Also, buildings and pipelines disrupt the natural migration paths of animals like caribou **b**. Disposal of waste generated by workers causes further problems, as making landfill sites on permafrost **a** is difficult and low temperatures slow down decomposition **b**. All this shows that even when developments in cold environments try to be as sustainable as possible, there is often lasting damage to the environments and ecosystems.

e **7/9 marks awarded** **a** The answer is relevant, concise and accurate and uses a place-specific example (Alaska). **b** It successfully outlines several problems associated with economic development in cold environments. **c** Some reference (though limited) is made to Figure 6. **d** In places there is digression into the solutions to environmental problems. The answer could be improved by: the inclusion of a brief description of the unique challenges facing developments in cold environments and the fragility of cold environment ecosystems; a little more amplification of problems (at the expense of management); and consideration of general economic and demographic issues such as isolation and low population densities. Despite these caveats, the answer just achieves Level 3.

e **The total score for this four-part answer is 17/25, a solid grade C.**

A-grade answer

(a) (i) Mass balance refers to a glacier's budget **a**. Glaciers are stores of ice which receive inputs of fresh snow and outputs of meltwater and water vapour **b**. During the year the upper parts of the glacier have a net gain of snow and ice **a**. This is the accumulation zone. Meanwhile the lower parts have a deficit. This is the ablation zone **a**. The location on the glacier where the annual inputs of snow and ice are exactly balanced by outputs of meltwater and vapour **b** is known as the equilibrium line **a**.

e **4/4 marks awarded** **a** The answer is accurate and refers to the three main features shown in Figure 5 — accumulation zone, ablation zone and equilibrium line. **b** There is sufficient detail and relevance to award a top Level 2 score.

(ii) If there was an increase in snowfall, but no change in the rate of ablation, the mass of ice in the glacier would increase **a**. The result might be an increase in the thickness of the ice and/or an advance of the ice front **b**. Such changes are due to a lowering of average temperatures or an increase in precipitation **a**. If the conditions were reversed, with either an increase in temperature and/or a reduction in precipitation, the glacier would shrink and retreat up its valley **a**.

Glaciers advance during a glacial period **b** or ice age, and retreat during a warmer inter-glacial **b** or inter-stadial period **a,b**. When glaciers advance they erode the landscape, deepening U-shaped valleys and transporting moraine. During a period of retreat, glacial deposition dominates, and glaciers leave behind features such as terminal moraines, recessional moraines and lateral moraines **c**.

5/6 marks awarded **a** The link between glacial processes and glacier behaviour is clear and is fully explained and relevant. **b** Terminology is used correctly throughout. **c** The last section on glacial features is not strictly relevant. Although marks are not deducted for the diversion into glacial landforms, an opportunity to provide additional relevant detail is missed. Despite some irrelevance this is a good answer, which achieves Level 2.

(b) Moraine is an unsorted mixture of rock debris, varying in size from boulders to fine rock flour **a**, and deposited by a glacier **b**. Three common morainic features found in glaciated lowlands are terminal moraines, drumlins and till plains **a**. Terminal moraines form when the front of a glacier or ice sheet is stationary for an extended period of time **c**. Moraine carried forward by the ice builds up at the ice front **a**. Eventually as the ice melts or the glacier retreats, the moraine forms a ridge of low hills that may extend for many kilometres across the landscape **c**.

Drumlins are small, streamlined hills made of moraine **c**. In plan they have an oval form, and in cross section they have a short steeper slope (stoss) and a longer gentler slope (tail) **a**. Drumlins are formed by moving ice beneath an ice sheet. Their streamlined shape suggests they have been moulded by moving ice.

Till plains are vast sheets of moraine that plaster entire landscapes **c**. Like drumlins they are formed by ice sheet deposition. Sometimes they consist of ablation till **a** (deposited by the melting of stagnant ice). On other occasions the moraines are deposited by moving ice. This deposit is known as lodgement till **a,d**.

5/6 marks awarded **a** Terminology is used accurately, and conveys a secure understanding of the topic. **b** The answer starts sensibly by defining the term moraine. **c** Although the answer is confined to just three types of moraine (lateral, medial, recessional, with ground moraines etc. being ignored), these features are described in detail and the explanations are sound. **d** The answer recognises the contribution of ice sheets to the formation of moraines. This again suggests an impressive depth of understanding. Some references to actual examples and a greater range of morainic features would strengthen the answer. Nonetheless, the answer achieves Level 2.

(c) Figure 6 shows that in the past 20 years the temperature of the ground has increased, threatening to melt the permafrost **a**. Permafrost melting creates massive problems for development in the arctic and sub-arctic. Widespread subsidence occurs when the permafrost melts, damaging buildings and infrastructure such as roads, railways and pipelines **b**.

Low temperatures and the fragile ecosystems of cold environments add to the problems of development. In Alaska, oil and gas companies have to cope with sub-zero temperatures for 8 months of the year. Oil spillages at low temperatures result in long-lasting damage to the environment, with slow rates of organic decomposition **b**. For the same reason disposing of sewage is also difficult **b**.

Damage to the fragile vegetation cover of mosses and dwarf shrubs takes many years to recover and at the same time exposes the permafrost to melting **b**.

In Alaska the building of the Trans-Alaskan Pipeline has disrupted the migration of caribou to their summer calving grounds on the North Slope **b**. This in turn threatens the livelihoods of native people like the Inupiat who depend on hunting **b**. Exploration for oil offshore in the Arctic Ocean using seismic surveys through controlled explosions scares away animals such as seals and whales on which native hunters depend **b**.

Although some of these problems are not unique to cold environments, the fragility of ecosystems (including short food webs, low biodiversity and low productivity) makes these problems especially acute.

e **8/9 marks awarded** **a** The answer makes clear reference to Figure 6. **b** The emphasis on problems related to the tundra is relevant and well exemplified with place specific detail on Alaska. Discussion and explanation of environmental problems shows sound knowledge and understanding. Overall, the answer has sufficient depth, relevance and exemplification to reach Level 3. However, there is scope for some discussion of non-environmental problems, and references to other types of cold environments (e.g. glaciated uplands).

e **The total score for this four-part answer is 22/25, an A grade.**

Question 4 **Hot arid and semi-arid environments**

Table 1 Climate at Riyadh, Saudi Arabia

Month	Mean minimum temperature (°C)	Mean maximum temperature (°C)	Mean precipitation (mm)
January	8	21	3
February	9	23	20
March	13	28	23
April	18	32	25
May	22	38	10
June	25	42	0
July	26	42	0
August	24	42	0
September	22	39	0
October	16	34	0
November	13	29	0
December	9	21	0

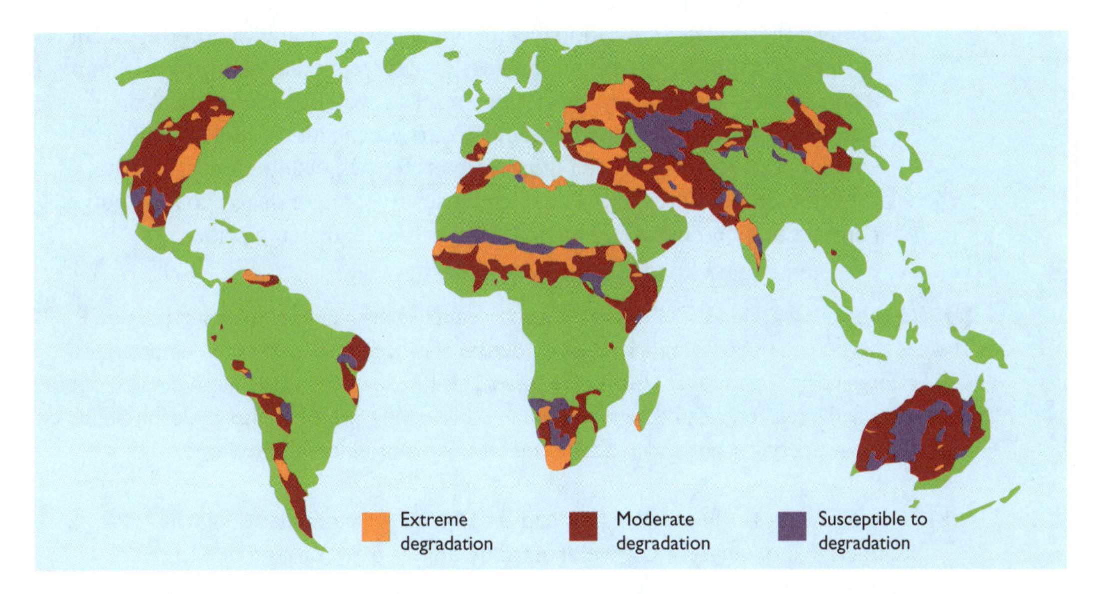

Figure 7 Land degradation in the world's drylands (source: UNEP)

(a) (i) Use Table 1 to describe the main features of the climate in Riyadh in Saudi Arabia. (4 marks)

(ii) Outline the weathering processes that are likely to occur in hot arid and semi-arid environments. (6 marks)

(b) Explain how running water has influenced the development of landforms in hot arid and semi-arid environments. (6 marks)

(c) With the aid of Figure 7, and with reference to one or more located examples, explain how human activities can often cause irreversible damage to hot arid and semi-arid environments. (9 marks)

ⓔ Good answers to (a)(i) will cover seasonal trends in monthly maximum and minimum temperatures, annual range of temperature, and monthly and annual rainfall. Answers will be illustrated with actual values from Table 1. In (a)(ii) the command 'outline' is asking for a brief description. In (b) and (c) the command 'explain' must be followed explicitly. Lapses into description must be avoided.

C-grade answer

(a) (i) Riyadh's hottest months are June, July and August, all having a mean maximum temperature of 42°C **a**. The coldest month is January with a low of 8°C **a**. Only 3 months have 20mm or more precipitation — February, March and April — with the maximum being 25mm in April. June to December sees no precipitation on average.

ⓔ **2/4 marks awarded** **a** Although accurate, this answer is simply a listing of the main features. This is a Level 1 answer. It is brief and ignores some important features of Riyadh's climate, such as the large mean annual range of temperature and the mean monthly temperature range. Some attempt could have been made to describe the yearly precipitation pattern.

(ii) There are three main types of weathering processes: freeze–thaw weathering, salt weathering and insolation weathering **a**. Freeze–thaw weathering occurs in semi-arid mid-latitude environments. Water trapped in confined rock joints **b** and crevices expands by 9% on freezing. The pressure exerted by the ice shatters **b** the rock **c**. Salt weathering occurs when salt crystallises out of solution **b**. The growth of salt crystals leads to rock disintegration **b,c**. The range of temperatures in hot deserts causes rock minerals to expand and contract at different rates which also leads to rock disintegration **c**.

ⓔ **4/6 marks awarded** **a** The answer correctly identifies the main weathering processes. **b** There is appropriate use of terminology. **c** The outline of processes is sound and demonstrates solid understanding. This answer reaches the top of Level 1. However, a little more detail about salt weathering and insolation weathering is needed to achieve the next level. Also, the contribution of moisture to weathering in hot arid and semi-arid environments is not fully explored.

(b) Running water is the process that causes alluvial fans, bajadas **a**, playas, pediments and canyons to develop **b**. Alluvial fans **b** are formed due to rivers quickly losing energy in lowlands, resulting in them depositing sediments across a broad area **c**. Many streams and rivers in hot arid and semi-arid environments drain to shallow inland basins, where they form playas **b,c**. Pediments **d** are formed by the parallel retreat of steep slopes of plateaus or mountain fronts. Parallel retreat occurs because rock debris is removed as fast as it accumulates. Running water is an agent of both transport and erosion. Canyons are the result of vertical erosion by running water. Abrasion due to the scouring action of coarse rock debris in transport is the dominant process **c**.

4/6 marks awarded **a** Bajadas are listed, but not explained. **b** The answer considers a range of landforms that owe their development to running water. A brief description of each landform would enhance the answer. **c** There is only limited explanation and insufficient emphasis on the relationship between process and form. **d** It is arguable that pediments are the outcome of weathering rather than fluvial processes. The power of running water in desert environments and its ability to erode landforms such as canyons has been ignored. Some reference to changing climate and the wetter conditions that prevailed in the past in many desert environments would help to give a fuller picture. This is a Level 1 answer: it is not fully developed and has significant omissions.

(c) Looking at the map one can see that the more populated areas of hot arid and semi-arid environments have more extreme degradation. Even slight disruption can cascade through desert ecosystems and cause irreversible change. The more populated the area, the more extreme the damage. For example, areas that attract a lot of tourism such as the Arches National Park have cars parked along road shoulders which damage soil and vegetation. Human activities also damage the fragile cryptobiotic crusts **a**, which take 50 to 250 years to recover **b**. Cryptobiotic crusts are easily crushed and damaged by vehicles, mountain bikes, hikers and livestock as they are thin and brittle **b**. In the Arches National Park, bikers and walkers who follow popular trails damage the cryptobiotic crust. Damaged crusts contribute less nitrogen and organic matter to the soils, and are more likely to be blown away by the wind. This in turn exposes soils to erosion **a**. Blown soil then buries nearby crusts, stopping photosynthesis **a** and causing further damage **b**.

6/9 marks awarded **a** The use of geographical terminology, grammar and spelling is good. **b** The fragility of the environment and slow rates of recovery from damage caused by human activities is explained clearly and in some detail in the Arches example. Overall this is a good Level 2 answer. However, its focus on a single problem (damage to cryptobiotic crust) makes it rather narrow. More reference to Figure 7 and a slightly broader approach, with some explanation of the fragility of hot arid and semi-arid environments, are needed to achieve Level 3.

The total score for this four-part answer is 16/25, a sound C grade.

A-grade answer

(a) (i) Riyadh's temperatures vary from warm during winter to hot in summer **a**. Even the coldest month (January) has a maximum monthly temperature of 21°C **b**. By July, average maximum temperatures have soared to 42°C **b**. In winter, night time minimum temperatures drop to an average of 8°C in January **b**. Mean annual rainfall is just 81mm **b**. Rainfall is concentrated between February and April **a**, and no rain falls between June and December **a**.

4/4 marks awarded **a** This answer clearly identifies the main temperature and rainfall patterns. **b** The temperature and rainfall patterns are supported with climatic data from Table 1. The answer is descriptive, and avoids straying into irrelevant explanation. This is a top Level 2 response.

(ii) The main weathering processes in hot arid and semi-arid environments are salt weathering, insolation weathering and freeze–thaw weathering **a**. Salt weathering happens when salt in solution is absorbed by permeable rocks. As the rocks dry out salt crystals grow out of solution. They exert pressure on the rocks causing them to disintegrate. Often salt weathering causes a characteristic surface flaking of rocks **b**.

Insolation weathering occurs when surface rocks are heated during the summer. Surface temperatures may reach 80°C. The high temperatures cause minerals in the rocks to expand and contract at different rates. With the presence of some moisture (from dew or rain) rocks are weakened and start to break down **b**. Again weathering is concentrated near the surface. The surface rock layers may peel away in a process called exfoliation.

Finally in hot arid and semi-arid environments at altitude (e.g. Namibia) clear skies in winter may lead to frost **c**. Freeze–thaw weathering can occur even within the tropics. The growth of ice in cracks in the rocks exerts tremendous pressure, causing them to break up.

e 6/6 marks awarded a The answer shows good understanding of three weathering processes found in hot arid and semi-arid environments. **b** The answer is clear and detailed. For example, it recognises the importance of salt weathering and the role of water in insolation weathering. **c** The observation that freeze–thaw weathering can occur at altitude in the tropics is further evidence of a quality answer. This is a top Level 2 response.

(b) A number of landforms in hot arid and semi-arid environments owe their development to the action of running water. Features such as canyons, alluvial fans and playas **a** have been formed by rivers. However, some of these features may have formed thousands of years ago when the climate was much wetter than today **b**. Although rainfall is sparse in hot arid and semi-arid climates, rain does occur. Often rainfall is torrential, and for the short time that rivers flow they have massive power. Lack of soil and vegetation cover also means that rivers can transport large amounts of sediment.

Canyons are steep, narrow valleys with almost vertical sides. They are eroded in solid rock by fluvial action. Many canyons are still actively forming today, as the canyons in places like the Colorado Plateau and the Anti-Atlas mountains **b** in north Africa experience flash floods after heavy rain **c**.

Alluvial fans are cone shaped deposits of sediment found at the foot of mountains. They develop where a river emerges into lowland after being confined to a steep narrow valley in the mountains. As the river leaves the mountains there is a sudden loss of energy (it is no longer confined and the gradient is less steep) and it deposits its sediment load **c**.

Playas are shallow lakes or lake beds in arid environments which occupy shallow basins. They are the centres of inland drainage **c**. After rain they may fill with water, but the water soon evaporates and leaves behind deposits of salt. A good example of a salt-covered playa is found at Badwater in Death Valley **b**.

6/6 marks awarded **a** The answer correctly identifies three landforms and makes clear their connection with fluvial processes. **b** There is recognition that many landforms attributed to fluvial processes may have formed during past 'pluvial' or wetter conditions. **c** The answer is well structured, with a short explanatory paragraph devoted to each landform. It also includes relevant examples. This is a comprehensive answer that shows detailed understanding of the role of running water in hot arid and semi-arid environments. It is at the top of Level 2.

(c) Hot arid and semi-arid environments are fragile and easily damaged by human activity. Often this damage is irreversible. Figure 7 shows that most of the world's drylands have suffered either extreme or moderate degradation **a**. The rest are susceptible to degradation. Land degradation describes the deterioration of land due to processes such as soil erosion, desertification and salinisation **b**.

In the Korqin Sandy Lands of China **c**, large-scale soil erosion has occurred due to poor farming methods and population pressure **b**. Rising population has caused local farmers to over-exploit soil and vegetation resources. Domestic animals such as goats, cattle, camels and horses have overgrazed **b** the land and destroyed the vegetation cover. The problem has been made worse by deliberate clearing of woodland for farming and timber. Increased demands for water for irrigation and industry have lowered the water table **b** resulting in soils drying out. Finally, a shortage of manure has reduced the soil's organic content. The outcome is fragile, over-cultivated and exhausted soils. These soils are easily eroded and desertified. Wind erosion **b** is common and creates violent dust storms. The storms blow the topsoil away. Reversing desertification will not be easy, especially as it has occurred over thousands of square kilometres **d**.

Hot arid and semi-arid environments are also susceptible to salinisation. Poor management of irrigation and drainage often leads to over-watering of crops and waterlogged soils **b**. In a hot climate, evaporation draws this water to the surface, where it leaves a deposit of salt. Eventually soils become so saline that farming has to be abandoned. This happened at Khushab in northern Pakistan **c**. By the late 1980s salinisation (due to over-irrigation) had become so bad that crop yields were down, land had been abandoned and there was high unemployment and poverty **d**.

Because hot arid and semi-arid environments are so fragile, with slow rates of regeneration **b** of soil and vegetation resources, damage to the environment caused by poor management is often irreversible.

9/9 marks awarded **a** The answer adheres to the question and makes references to Figure 7. **b** The answer is well structured, relevant and uses geographical terminology accurately. **c** Description and explanation are supported by two place-specific examples. **d** A range of relevant and detailed causes and effects are discussed. Throughout the answer, the focus is exclusively on problems; there is no slippage into irrelevant discussion of management and solutions. This is an excellent answer, which achieves the top of Level 3.

The total score for this four-part answer is 25/25 marks, an outstanding A grade.

Section B: extended-writing questions

Question 5 **River environments**

With reference to located examples, explain how management might help to resolve conflicts between development and flood-risk issues. (25 marks)

e The crucial commands in this question are 'explain how' and the utmost care is needed to avoid lengthy (and irrelevant) descriptions of conflicts between development and flooding. Good answers will also use at least two detailed examples. As with all extended answers, always start with a brief introduction (definition of terms, outline of the structure of the discussion that is to follow), and end with a short conclusion (a summary of the main arguments, a reiteration and justification of your viewpoint/opinion if needed).

C-grade answer

The Mississippi and Brahmaputra rivers have been managed to some degree to control flooding. Investment in flood control has been greater on the Mississippi, with thousands of kilometres of levées, over 200 dams, meanders straightened and wing dykes built **a**. However, river engineering has not stopped flooding, as the 1993 and 2008 floods showed. The main issue between development and flooding is whether housing, factories, offices, roads etc. should be built on floodplains. After the 1993 Mississippi floods many planners argued that floodplains should be zoned solely for land uses that were not vulnerable to flooding such as pastureland, sports fields and so on **a** But with much of the Mississippi floodplain already built on, management has to centre on ways of strengthening flood defences and preventing floods from developing in the first place (e.g. land use control in the headwater regions) **a**,**c** There is no doubt that levées will continue to be maintained and strengthened. However, there is only so much that can be done. It is unrealistic to expect a major relocation away from the floodplain, especially when the Mississippi flows through so many major cities (e.g. St Louis, New Orleans). People on the floodplain will just have to learn to live with floods.

Attempts are being made to 'tame' the Brahmaputra River in Bangladesh, which floods every year due to the monsoon and snowmelt in the Himalayas **c**. Flooding in Bangladesh does not just damage property. It kills thousands of people, paralyses the country's economy and spreads disease. There are ambitious plans to control the Brahmaputra and protect people and economic activities on the floodplain **b**. If the Brahmaputra could be controlled it would give a big boost to Bangladesh's economic development. The current management schemes involve straightening the river's course and building massive levées **a**. But the Brahmaputra is a more powerful river than the Mississippi and Bangladesh is an overcrowded and poor country. So, given the scale of the task and the country's limited resources, there seems little likelihood of any major advance in flood control in the foreseeable future.

17/25 marks awarded **a** The answer makes appropriate references to hard engineering solutions to flood management; soft engineering through land use change and flood abatement; and the issue of land use in flood-risk areas. **b** In terms of content there is some lack of detail (only around 350 words in length). For example, flood control schemes planned for the Brahmaputra could have been developed further. **c** Overall, the answer is well structured with accurate spelling, grammar and terminology. Contrasting examples of the Mississippi and Brahmaputra are used effectively and the answer is pleasingly analytical and focuses on the issues. Cause and effect are well understood and there is the recognition that there will always be conflict between flooding and development. However, the absence of a conclusion places it in Level 2 rather than Level 3.

The answer scores 9 marks for knowledge and understanding, 3 marks for analysis and application, and 5 marks for skills and communication, to give a high C grade.

A-grade answer

Development often conflicts with flood-risk issues because development can on the one hand increase the frequency and size of floods, and on the other can increase the number of properties that are vulnerable to flooding.

Development can increase the frequency and size of floods by changing land use. For example, development that causes deforestation, increases urban growth and converts arable land to pasture, is likely to increase the risk of flooding. The reason for this is that the amount of rainfall that becomes runoff increases. This may be due to a reduction in water storage (e.g. in soil and vegetation) and evapotranspiration. When water is stored naturally it is released slowly, reducing peak flows and minimising the risk of flooding **a**. Evapotranspiration simply removes water that might have flowed into streams and rivers into the atmosphere.

Flooding has been a major problem in the Santa Ynez drainage basin near Santa Barbara in southern California. The main cause has been the mismanagement of the upper parts of the basin including deforestation, and the misuse of farmland that has caused widespread soil erosion as well as flooding **a**. Floods have damaged houses, railways, roads and farmland. Management that will help to prevent flooding includes encouraging farmers to grow cover crops, terracing fields and establishing vegetation in critical areas. Fire prevention is also vital to maintaining the forest cover. The fire service helps local people maintain firebreaks and fire lanes **a**.

In England, 1.7 million homes and 300,000 commercial properties are at risk of flooding. This is mainly because development has been allowed to take place on floodplains. Also, floodplains, under natural conditions, are important temporary stores of floodwater **a**. Draining floodplains and preventing rivers from flooding across these areas increases the flood risk further downstream **c**. Today planners in the UK aim to restrict new development on floodplains **b**. Where development is allowed, developers must assess the flood risks and take measures to reduce them. This can include building flood defences and maintaining them **b**. Even so, in a densely populated country like the UK, the pressure to develop floodplains is immense.

Selby, located on the River Ouse in North Yorkshire, has suffered serious flooding in recent years. Between 2004 and 2008 an ambitious flood protection scheme was completed, involving the construction of new embankments and the raising of existing ones along an 8.6 km stretch of the River Ouse, and a flood gate between the Selby canal and the river **b**. This shows how hard engineering can help to resolve the conflict between development and the flood risk **c**.

21/25 marks awarded **a** Content and understanding are detailed and cause and effect are well understood. There is some analysis of how management can help to resolve conflicts. **b** The answer provides a number of relevant arguments which are supported with convincing examples from California and the UK (Selby). **c** The answer is well structured with accurate use of grammar, spelling and geographical terminology. However, the absence of a conclusion places the answer in Level 2 for skills and communication.

This is a relevant answer, which addresses the question effectively. It scores 12 marks for knowledge and understanding, 4 for analysis and 5 for skills and communication, to give a marginal A grade.

Question 6 **Coastal environments**

With reference to located examples, explain how management is needed to resolve development issues and conflicts in coastal environments. (25 marks)

e A good answer will focus on two areas: an outline of some development issues and conflicts; and an explanation of how management has helped to resolve them. It is essential that discussion is within the context of at least two detailed examples.

B-grade answer

The Holderness coastline in east Yorkshire **a** is the most rapidly retreating stretch of coastline in Europe, retreating on average 2m a year. The coast consists of soft, easily eroded boulder clay **a**. Rapid erosion is also due to the absence of beaches so there is nothing to protect the cliffs. As the coast is retreating so rapidly, the homes and businesses of people who live there are threatened **a**. Over 30 villages have been lost on the Holderness coast since Roman times. Local people want something to be done to protect them against erosion **a**.

At Mappleton, three types of hard defence have been used to stop erosion. They have saved the village and the main road between Hornsea and Withernsea from erosion **b**. Two large rock groynes stop longshore drift and have built a wide beach which absorbs wave energy **a**. Armour blocks placed at the base of the cliffs protect the cliffs from erosion. Meanwhile the cliffs have been regraded to a lower angle so that rotational slides no longer occur. Since these measures were taken in the early 1990s, coastal erosion has been halted **a**,**c**.

However, just south of the new defences, rates of erosion on the unprotected coast have accelerated. Farmland and cliff-top farms are at greater risk **c**. One farm at Great Cowden had to be demolished in 1998 before it fell into the sea **a**. It is argued that the defences at Mappleton have increased erosion further down the coast **b**. This is because the rock groynes intercept sand and shingle moving south along the coast. As a result beaches downdrift are starved of sediment and become thin. Also, stopping erosion at Mappleton means that less sediment gets into the sea to build beaches **a**.

This example shows how management is needed to plan the future of the whole coastline **c**. Solving a problem in one place can create problems elsewhere **c**. Shoreline Management Plans (SMPs) are designed to tackle this situation. They consider long stretches of coastline and plan them as a whole. Today Holderness is part of the SMP that stretches from Flamborough Head in east Yorkshire to Gibraltar Point in Lincolnshire. The SMP maps the long-term future of the coastline based on principles of sustainability.

e **18/25 marks awarded** **a** This answer provides some detailed knowledge and is place-specific. **b** Knowledge and understanding are applied relevantly to the question, and cause and effect are well understood. The main weakness is the answer's limited scope, which concentrates exclusively on coastal erosion and uses only a single example. Other development issues that could have been considered include: managed retreat/managed realignment; offshore mining of sand; the construction of offshore wind farms; building links golf courses in coastal dune environments. As a result the answer cannot achieve more than Level 2 for knowledge and understanding. **c** There is

some analysis of how management can help to resolve development issues, which suggests Level 2 for analysis and application. The answer is well structured, with good use of grammar and spelling, and has a conclusion. For skills and communication it reaches Level 3.

e Overall the answer scores 9 marks for knowledge and understanding, 3 for analysis and 6 for skills and communication, to give a low B grade.

A-grade answer

Conflict often arises when development occurs in coastal environments. Development can damage fragile coastal environments such as sand dunes, ruin coastal landscapes through unplanned buildings, pollute coastal waters and increase flood risks **a**.

Sand dunes support fragile ecosystems that are easily damaged. However, dunes are attractive to recreation and tourism and some are put under severe pressure. If the surface vegetation cover of dunes is destroyed, they are easily eroded by the wind. The largest area of dunes in England is on Merseyside, between Liverpool and Southport **b**. These dunes are easily accessible and are a major attraction to visitors from nearby towns and cities. Although the most sensitive areas are protected as nature reserves, elsewhere trampling, firing, littering and overuse has caused massive erosion and blow-outs. Quad biking and motorcycling add further to the erosion problem **b**. Apart from their importance as ecosystems, the dunes play a significant part in coastal defences, helping to protect areas inland from flooding. Careful management that restricts access and provides boardwalks in the most popular areas are needed, as well as plans for the re-establishment of vegetation and the stabilisation of dunes **c**.

On the Costa del Sol in southern Spain, much of the coastline has been degraded by unplanned urban development associated with the growth of tourism **b**. Hotels and apartments line the coast for mile after mile; buildings come to within 20 m of the shoreline; and pollution of beaches by untreated sewage effluent is still widespread. Meanwhile, large numbers of golf courses not only destroy the natural environment but also consume huge quantities of water. Overdevelopment **d** threatens the future of tourism as visitors, dissatisfied with the poor quality of the environment, holiday elsewhere. Modern planning aims to restrict new urban development and remove some of the worst eyesores built in the 1970s and 1980s **c**. Many houses and hotels, built without planning permission, have been demolished in recent years. Restrictions are being placed on the development of new golf courses, and new sewage works are being built and existing ones expanded **b**.

Conflicts can arise when coastal defence works, designed to stop erosion and flooding, resolve problems on one stretch of coast but create problems elsewhere. Groynes and sea walls interfere with the movement and input of sediment to the coastal system. These structures are blamed for accelerated erosion along parts of the Holderness (between Mappleton and Withernsea) and northeast Norfolk (around Happisburgh) coasts. These problems could have been avoided by coastal management **c**. Today, all stretches of coastline in England are covered by Shoreline Management Plans (SMPs) **d**. These plans are based on detailed studies of the coastal system and local sediment budgets **d**. Decisions to protect the coast in one area are considered in relation to the effects on the entire system.

Finally, in the delta region of Bangladesh **b**, development has caused massive pollution and destroyed extensive areas of mangrove. Pollution of coastal waters with toxic chemicals from industry and intensive agriculture accumulates in fish and other marine animals and gets into the human food chain. Untreated sewage causes eutrophication **d** in coastal waters and damages fish stocks. Destruction of the mangroves, due to population pressure and poverty, and the development of shrimp farming have exposed coastal settlements to greater risks of flooding by tropical cyclones and tidal surges **d**. These problems are being tackled by the Bangladeshi government's integrated coastal zone management plan **c**, which aims to achieve a more sustainable use of resources and protect ecosystems, while reducing poverty and the exposure of people to flooding in the delta region **b**.

In conclusion, it is clear that development issues in coastal environments can only be resolved through management. When management is in place (e.g. England's SMPs) conflicts can be resolved. Without management, the free-for-all approach between different activities leads not only to conflict but to environmental degradation, which ultimately can put lives and property at risk **b**.

e **24/25 marks awarded** **a** The answer has a clear structure which is flagged in the introduction. Spelling and grammar are secure, and there is logical conclusion. **b** There is an impressive amount of detailed knowledge and understanding, supported by relevant examples. **c** Analysis of situations where management is needed to resolve conflicts in coastal environments is clearly targeted and relevant. **d** There is varied and accurate use of geographical terminology. This is a detailed and lengthy account written around a number of place-specific examples.

e **The answer scores 13 marks for knowledge and understanding, 4 for analysis and 7 for skills and communication, to give a top A grade.**

Question 7 **Cold environments**

With reference to located examples, explain how cold environments can be managed to ensure their sustainability. (25 marks)

e Start with definitions of cold environments and sustainability. The link between management and its sustainability must be emphasised throughout. This must be done within the framework of located examples.

B-grade answer

It is important to ensure sustainability in any environment. But cold environments have an extremely fragile ecosystem and structure, so extra care is needed to prevent lasting damage.

In the Himalayas in Nepal **a** there is fast population growth, poverty, deforestation and uncontrolled tourism. This has resulted in a number of environmental and social problems. Local people have been swamped by the number of tourists and trekkers burning trees for firewood and creating litter. Visitors have also caused problems of trail erosion and have trampled vegetation and damaged animal habitats. Habitats have been further damaged by overgrazing by livestock owned by subsistence farmers, leading to deforestation **b**.

These problems are being tackled to make tourism and local farming more sustainable. Nepal has over 13,000km^2 of protected areas including eight national parks. The Annapurna Conservation Area (ACA) covers a large area and includes the world's eighth highest mountain. The area attracts large numbers of ecotourists who come to experience the mountains, waterfalls and wildlife. The ACA project was set up to protect wildlife and biodiversity. Tourists are charged $7 to enter the area, and this, together with funding from the World Wide Fund for Nature and the US government, provides a source of income for local people **a**. Local people are employed to protect the environment. The ACA supports jobs for local people such as tree planting, and wildlife, water and soil conservation **a**. Local people are also employed as forest guards, tour guides, and in visitor centres and mountain rescue teams. Local people have a stake in conserving the environment that at the same time helps them sustain their way of life. Education has led to better management of livestock and re-planting trees **a**.

On the North Slope of Alaska **a**, the extraction of oil threatens the environment, damaging vegetation, disrupting animal migrations, causing the permafrost to thaw and destroying the traditional way of life of native Indians. However, attempts are being made to manage the environment sustainably using 'environmentally friendly' technology. This includes seismic exploration **b** to reduce the number of drilling sites, building roads on insulated ice pads to prevent the permafrost from melting, using computers to pinpoint oil and gas-bearing rocks, and drilling laterally **b** to avoid the use of drilling rigs in the most environmentally sensitive areas **a**,**b**.

18/25 marks awarded **a** The answer refers to two specific examples and provides some detail of the management strategies for sustainable development. **b** Knowledge and understanding for a 25-mark question needs to be more detailed (especially in the Alaskan example) to access Level 3. The answer is generally well written, with effective use of grammar and spelling. The content is accurate and for the most part relevant. However, the account could have been improved by providing some evaluation of management strategies in the two examples, a definition of the concept of sustainability in the opening paragraph, and a brief conclusion of no more than five or six lines.

The answer scores 9 marks for knowledge and understanding, 3 for analysis and application and 6 for skills and communication, to give a low B grade.

A-grade answer

Cold environments with their low biodiversity, limited plant cover and slow rates of plant growth, are easily degraded by human activities. Cold environments with permafrost rely on a delicate thermal balance. If this is disrupted it can cause irreversible environmental damage.

In Alaska **a**, the exploration and exploitation of oil and gas reserves on the North Slope has often been unsustainable and has caused widespread, though localised, environmental damage. The biggest current threat is posed by oil exploration in the Arctic National Wildlife Refuge (ANWR).

It is hoped that new technologies will limit the environmental footprint of the oil and gas industry in the ANWR. Seismic exploration means that less drilling is needed and so fewer access roads and other infrastructures are built on the tundra. Roads and pipelines are built on insulated ice pads to protect the permafrost, while powerful computers can help to locate new oil and gas fields, reducing the impact on the environment. Finally, new lateral drilling methods allow oil and gas to be accessed several kilometres away from drilling rigs **b**.

But despite these developments, new technologies cannot eliminate the impact of the oil and gas industries completely. Occasionally oil spillages will occur, gravel will have to be extracted from river beds to make insulating pads, and airstrips and roads will be needed. Migratory movements of caribou, on which local indigenous people depend, will still be disrupted **b**.

In the Annapurna region of the Himalayas in Nepal **a**, poverty and population growth has caused deforestation and environmental degradation. This problem has been compounded by the effects of tourism. Trekkers exploit the forests for firewood, and pack animals overgraze areas around trails, many of which have been badly eroded. Waste generated by trekkers pollutes streams and rivers, and trash discarded by trekkers litters the environment **b**.

In an effort to manage the area's resources sustainably, the Nepalese government has created the Annapurna Conservation Area (ACA). The management plan is community-based and involves local people. It aims to protect wildlife and maintain biodiversity, and improve the quality of life for local people. Tourists are charged a fee to enter the ACA and this money funds conservation programmes that provide jobs for local people. The work involves mending eroded trails, planting trees, improving

sanitation facilities and acting as forest guards. Recent surveys show that since the introduction of the ACA the forest area has increased and so too has biodiversity. As local people have an additional source of income, pressure on the environment is reduced. The scheme has also changed local people's attitude towards the environment. They now see that it is in their interests to conserve the environment, the resource that tourism ultimately depends on **b**.

However, conservation by itself is not enough to ensure sustainable development. Local people will have to reduce their dependence on fuelwood and rely more on kerosene — and the problem of rapid population increase will have to be tackled.

e **22/25 marks awarded** **a** This answer is closely based on two contrasting case studies. This approach ensures that the answer is place-specific and avoids unnecessary generalisation. **b** There is good depth and accuracy to the case studies, which examine problems and solutions. Both case studies are appropriate and relate closely to the question. The overall level of detail and application of knowledge and understanding is good. For this, the answer achieves Level 3, and analysis is clearly Level 3. Although the answer is well structured, the lack of a conclusion reduces the answer to Level 2 for skills and communication.

e **The answer scores 12 marks for knowledge and understanding, 5 for analysis and application and 5 for skills and communication, to give an A grade.**

Question 8 **Hot arid and semi-arid environments**

With reference to located examples, explain how hot arid and semi-arid environments might be managed to ensure sustainability. (25 marks)

Definitions of hot arid and semi-arid environments and sustainability are needed at the outset. Simply describing management strategies in the context of chosen examples is not enough to merit a high grade. Successful answers will concentrate on demonstrating how management aims to/has achieved, sustainable outcomes.

E-grade answer

In the Arches National Park in Utah a, people travel from all over to see the popular attractions such as the Delicate Arch, Garden Wall and the Windows. At peak times the car parks reach full capacity so people park cars along road shoulders. This damages soil and vegetation b. To ensure sustainability the park plans to implement a new transport strategy c. This will include reserved parking at key attractions, closing roadside pull-off areas and controlling illegal parking b.

China's desert land suffers from desertification. It affects one-third of the country's total land area. The most seriously affected areas are the arid regions in northern China a. In 1978, China began an ambitious programme to tackle desertification d and ensure sustainability d. The programme is known as the 'great green wall' b. Its purpose is to establish 350,000km^2 of shelterbelt plantation forests d across the entire region of northern China. The objectives are to protect farmland and settlements from wind and water erosion d, to improve management, stabilise d sand dunes, and reclaim degraded lands c.

12/25 marks awarded **a** Both named examples are appropriate but lack details of management strategies. Although the answer does address the question, for 25 marks it is too brief. As a result, important content such as land drainage and water management on salinised farmland, fencing pastureland and controlling grazing and stocking density, strip cropping, terracing of slopes and restricting access by private cars to the most sensitive areas in national parks is ignored. The answer also lacks any definition of sustainability and fails to explain why sustainability is a particular issue in hot arid and semi-arid environments. **b** In terms of knowledge and understanding (maximum 13 marks) the answer is at the top of Level 1, providing some relevant (but sparse) content and appreciation of management responses. **c** The answer achieves Level 2 for analysis and application because, despite its brevity, it applies knowledge and understanding relevantly to the question. **d** The answer scores Level 2 for skills and communication. In the context of its brevity, the answer is well written, and uses geographical terminology accurately.

The answer scores 5 marks for knowledge and understanding, 3 for analysis and application and 4 for skills and communication, to give an E grade.

A-grade answer

Hot arid and semi-arid environments have frequently been degraded by human activity **a**. Overgrazing and the removal of trees for building and firewood have led to soil erosion and desertification in many places. Over-irrigation **a** has caused salinisation of soils, reduced crop yields and sometimes resulted in land being abandoned altogether. All these processes contribute to poverty among farmers in hot arid and semi-arid environments. In MEDCs, hot arid and semi-arid environments are sometimes over-used by tourism, damaging vegetation, creating soil erosion and causing lasting environmental damage.

Desertification is the degradation of land to the point where desert-like conditions prevail. It affects one-third of China's total land area. In China's northern provinces desertification is being tackled by a huge reafforestation **a** programme: 350,000km^2 of forest are being planted. The forest will protect farmland and settlements from wind and water erosion, improve land management, stabilise sand dunes **a** and reclaim degraded land **b**. In the Korqin region of northeast China, the reafforestation programme is not just about planting trees. New cultivation techniques have been introduced to local farmers such as recycling organic material to improve the soil's structure, integrating tree crops **a** with pasture and arable land, and planting trees that provide fodder crops and improve soil fertility **b**. The management of grazing land is being introduced for the first time. The programme will help make economic activity more sustainable **a**, but only in the long term.

In Kushab in northern Pakistan, management is concerned with reclaiming land degraded by salinisation due to over-irrigation **a**. The project is funded by the World Bank, and aims to reduce waterlogging and salinisation on 360km^2 of cultivated land **b**. If successful it will increase farm production, create jobs and raise farmers' incomes. Waterlogging and salinisation are tackled by underdraining **a** farmland with PVC pipes, constructing new drains, and lining irrigation canals to reduce seepage **b**. Despite the progress that has been made, the project demonstrates that where land has become salinised and waterlogged, reclamation is difficult and costly.

The management problems in the desert lands of Utah, in the Arches National Park, are quite different. There the main concern has been damage to fragile soils and vegetation caused by off-road parking, traffic congestion at the most popular locations, and damage by bikers and walkers to the cryptobiotic crust **b**. Sustainable management in the park involves maintaining the 60km of paved road and parking areas, repairing trails and providing a ranger service. The park's transport plan includes reserved parking at key attractions, closing roadside pull-off areas and controlling illegal parking **b**. Areas of damaged cryptobiotic **a** crust are fenced-off, allowing recovery. However, recovery in such a harsh environment is slow and can take up to 250 years **b**.

In conclusion, hot arid and semi-arid environments easily suffer land degradation, and human activities are often unsustainable and contribute to poverty **a**. A variety of management plans are used, depending on the nature of the problems. All of them tend to be long term. In the meantime, rapid population growth often threatens to undermine these efforts.

24/25 marks awarded **a** This is well written answer, with accurate use of terminology, a purposeful introduction and a thoughtful conclusion. **b** It provides impressive and relevant detail of places and management strategies through its three case studies.

Overall this is an excellent answer, focusing on the question and applying knowledge and understanding appropriately. It is close to the top of Level 3 for all three assessment objectives. The answer scores 12 marks for knowledge and understanding, 5 for analysis and application and 7 for skills and communication, to give a very good A grade.

Knowledge check answers

1 Bedload comprises coarse sediments (cobbles, boulders), which slide and roll along the river bed during high-energy events (e.g. bankfull discharge). Bedload is usually transported over short distances.

2 Meandering channels are curving, single-thread channels, which develop where channel banks consist of coherent clay and silt. Braided channels are multi-thread, and interspersed with numerous gravel bars. Braiding is associated with unstable (easily eroded) gravel banks, high seasonal discharge and large, coarse sediment loads.

3
- Forestry — water supply: afforestation reducing water yield through increased interception and evapotranspiration.
- Water pollution — conservation/recreation and leisure/agriculture/water supply: pollution from agricultural runoff, effluent discharge by industry and sewage treatment and adverse impacts on other activities.
- Conservation — recreation and leisure: sensitive ecological sites protected exclusively for wildlife.
- Flood control/HEP generation (i.e. dams) — conservation/agriculture/settlement: impact of flood control on river regimes and wildlife habitats. Reservoirs drown farmland and settlements.

4 Physical reasons: intense rainfall; prolonged rainfall; rapid snowmelt. Human reasons: deforestation; urbanisation; artificial drainage of floodplains and wetlands; floodplain development. Note that floods are more likely to occur where catchments comprise impermeable rocks and have steep slopes and high drainage densities.

5 Advantages: low cost; conserves habitats and wildlife; is sustainable. Disadvantages: long timescales; complex land ownership issues (e.g. in reafforestation of upland catchments); economic and social disruption (e.g. restricting urban development of floodplains).

6 Weathering is the breakdown of rocks exposed at, or close to, the ground surface, by fluctuations in temperature and/or moisture levels. Mass movement is the downslope movement of rock and regolith (e.g. weathered material, soil) driven by gravity.

7 Rising sea level drowns valleys in coastal areas to create new landforms such as rias, fjords and estuaries. Wave action transports fluvial sediments, deposited on continental shelves during periods of low sea level, onshore, to form shingle and sand beaches.

8 Coastal protection based on hard engineering employs structures such as seawalls, groynes and revetments. These structures stop waves, obstruct longshore currents and raise the height of the coast. In this way they reduce rates of coastal erosion and risks of coastal flooding. They are costly to build and maintain and often have adverse and unforeseen effects on the coastal environment.

Soft engineering seeks to work with natural processes using techniques such as beach replenishment and managed realignment. It is environmentally friendly, relatively low cost and sustainable.

9 Potential conflicts between economic development and the physical environment in the coastal zone include: loss of wildlife habitat (e.g. salt marshes, dunes) to industry, transport etc.; overdevelopment of coastlines (e.g. by mass tourism), which damages the coastal environment; pollution of coastal waters by industrial effluent, farm effluent, sewage etc. and its impact on marine life, coral reefs etc.; deforestation (e.g. mangroves) increasing the risk of flooding.

10 Glacial moraines comprise weathered and eroded rock debris deposited directly by ice sheets and glaciers. The ice may be stagnant or moving and moraines may be transported on the surface, within or at the base of ice sheets and glaciers. Moraines are typically unsorted.

Glacio-fluvial deposits comprise sediments transported by meltwater. Deposition may occur on the surface on or beneath an ice cover, or beyond the ice front (pro-glacial). Unlike moraines, glacio-fluvial deposits are stratified (sorted by size into layers).

11 Climate influences tundra ecosystems by: low air temperatures limiting average net primary productivity; creating an extreme environment (short growing season, low solar radiation intensity, long periods of darkness in winter etc.), which reduces biodiversity and results in simple food webs; exposed conditions, which favour low-growing plants at the expense of trees; large year to year fluctuations in plant productivity, which have a knock-on effect on animal populations, creating cycles of 'boom and bust'.

12 Development often degrades tundra ecosystems because: melting of the permafrost disrupts drainage and causes slope instability and mass movement; lack of biodiversity means ecosystems have limited capacity to absorb change (implications for food webs and primary productivity); low temperatures result in the slow decomposition of organic materials (e.g. oil spillages); low temperatures and a short growing season produce slow rates of natural growth and regeneration.

13 Development in cold environments can be achieved sustainably by: minimising its impact on the physical environment and wildlife (e.g. conserving the permafrost layer, keeping development small-scale, accessing energy (e.g. oil and gas) remotely using new technologies); legislation to create large wilderness areas where development is prohibited; protecting indigenous societies and their cultures; confining building and construction to the winter months.

14 Abrasion is the scouring and erosion of rocks by sand particles entrained by the wind. Saltation is the transport of sand-sized particles by wind action; the sand, driven by the wind, skips and bounces along the desert surface. Deflation is the removal of particles finer than sand (e.g. clay, silt) by the wind. It is an important erosional process in deserts. Small, wind-borne particles may be transported thousands of kilometres.

15 Physiological adaptations to drought in desert environments include: water storage in plant tissues by succulents such as cacti; long tap roots to reach water deep underground (e.g. phreatophytes such as tamarisk); drought-evading plants that germinate and set seed in a few weeks following rain (e.g. desert paintbrush); plants that lie dormant during dry spells and spring to life when water becomes available — these plants often shed their leaves during drought or like the creosote bush have leaves with resinous coatings that reduce moisture loss; plants with needle-shaped leaves e.g. yucca.

16 Land degradation and desertification are caused by a combination of drought and unsustainable human activities. It is possible that climate change is increasing the frequency and intensity of drought. However, drought has always been a feature of arid and semi-arid environments. Over-exploitation of ecological resources (e.g. overgrazing, overcultivation, deforestation, falling water tables) resulting in soil erosion, loss of biodiversity and reduction in biomass is the main cause of land degradation and desertification.

17 Sustainable development does no long-term damage to natural resources such as climate, soil, water and ecosystems. It maintains biodiversity and ecological productivity and does not pollute and degrade the physical environment. It creates a balance whereby the rate of natural resource use is equalled by the rate of replacement, allowing resources to be exploited and recycled indefinitely. It also creates a stable environment, which is essential if the quality of life in MEDCs is to be maintained, and if living standards are to rise in the developing world in the twenty-first century.

A

B

C

D

E

F